CONSERVATION AND ECOLOGY

CONSERVATION AND ECOLOGY

Edited by John P. Rafferty, Associate Editor, Earth and Life Sciences

Britannica Educational Publishing
in association with
ROSEN
Educational Services

Published in 2011 by Britannica Educational Publishing
(a trademark of Encyclopædia Britannica, Inc.)
in association with Rosen Educational Services, LLC
29 East 21st Street, New York, NY 10010.

Distributed exclusively by Rosen Educational Services.
For a listing of additional Britannica Educational Publishing titles, call toll free (800) 237-9932.

First Edition

Britannica Educational Publishing
Michael I. Levy: Executive Editor
J.E. Luebering: Senior Manager
Marilyn L. Barton: Senior Coordinator, Production Control
Steven Bosco: Director, Editorial Technologies
Lisa S. Braucher: Senior Producer and Data Editor
Yvette Charboneau: Senior Copy Editor
Kathy Nakamura: Manager, Media Acquisition
John P. Rafferty: Associate Editor, Earth and Life Sciences

Rosen Educational Services
Heather M. Moore Niver: Rosen Editor
Nelson Sá: Art Director
Cindy Reiman: Photography Manager
Matthew Cauli: Designer, Cover Design
Introduction by Linda George

Library of Congress Cataloging-in-Publication Data

Conservation and ecology / edited by John P. Rafferty.
p. cm. — (The living earth)
"In association with Britannica Educational Publishing, Rosen Educational Services."
Includes bibliographical references and index.
ISBN 978-1-61530-307-6 (library binding)
1. Ecology. 2. Biodiversity conservation. 3. Nature conservation. I. Rafferty, John P.
QH541.C653 2011
333.95'16—dc22

2010017133

Manufactured in the United States of America

On the cover: Across the planet, only isolated sections in China's mountains remain habitable for giant pandas. Perhaps fewer than 2,500 survive in the wild. *Shutterstock.com*

On page x: El Capitan and the Merced River in Yosemite National Park, California. *Jeremy Woodhouse/Getty Images*

On pages v, 1, 37, 78, 101, 130, 184, 215, 234, 243, 245, 252: Giant panda (*Ailuropoda melanoleuca*). *Mike Clarke/AFP/Getty Images*

Contents

68

73

86

102

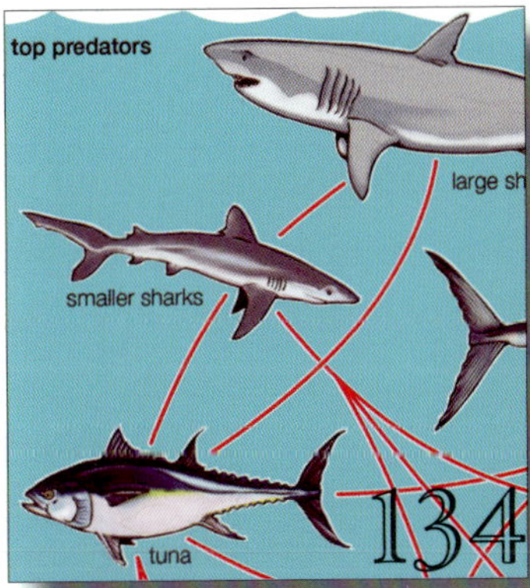
top predators
large sh
smaller sharks
tuna
134

145

160

162

178

219

226

INTRODUCTION

Conservation and ecology cover a vast array of topics related to the preservation of Earth's biosphere: the atmosphere, landmasses, and oceans where organisms, plants, and animals live. Species interact with one another within food chains and food webs as well as through other close relationships. These interactions can be disrupted by natural forces, such as climate changes, fire, and floods, as well as through the activities of human beings. Several challenges have focused the world's attention on Earth's environment and the conservation of natural resources. This book addresses those topics and how they relate to Earth's future as a living planet.

Although Earth's biosphere is primarily located on the planet's surface, it also extends some distance upward into the planet's atmosphere and downward into the planet's lithosphere. Ecology is the study of the relationships between living things and the environments they inhabit. The study of ecology includes the study of all forms of life, from the smallest bacteria to the largest animals and fungi. Ecology also considers the effects an organism's surroundings have on its survival and success.

Interacting species form biological communities. For example, a typical North American desert community is made up of coyotes, snakes, rabbits, birds, mice, various forms of cactus, creosote bushes, along with several species of insects, fungi, and other organisms. Communities of species interact with one another within an ecosystem, an ecological unit that includes the biological community along with its non-living surroundings. All species within a given ecosystem must contend with the demands made by the ecosystem's climate and topography. Different parts of an ecosystem are involved in nutrient cycling, which provides sustenance to the species that live there. Taken together, all ecosystems on the planet make up Earth's biosphere. To preserve the delicate balance inherent in ecosystems, biological communities, and populations of species, it is necessary to understand how organisms interact with one another.

Ecosystem function largely relies on the maintenance of nutrient cycles. Nutrients move through ecosystems driven by the actions of wind, water, and other physical forces, but they are also cycled through the organisms themselves through food chains and food webs. Food chains and food webs are structures that typically begin with plants. Energy from the sun is converted through a process called photosynthesis into plant tissues. Plant tissues (leaves, stems, roots, etc.) are consumed by animals called herbivores, which are, in turn, eaten by other animals called carnivores. Thus, energy passes through a chain of living players as food. Because most animals eat more than one plant or animal, however, a more complex structure called a food web emerges. Whenever a food chain or web is interrupted, through the sudden extinction of a species forming a link in the chain, the entire ecosystem may be changed. The degree of this alteration depends on how important the missing species is. In some cases, other species easily assume the role, or ecological niche, of the missing species. In others, where the species is an important food source or important predator, other species dependent on the missing species or its activities could become extinct as well.

Species also co-exist with one another symbiotically (that is, in close association). The interaction of two or more species can benefit each of the species involved (mutualism), harm both of the species involved (competition), benefit one species without benefiting or harming the other (commensalism), harm one species without benefiting or harming the other (ammensalism), or benefit one species while harming the other (predation or parasitism). Such close relationships are prevalent in ecosystems. In other situations when one species disappears from an ecosystem, the remaining species may adjust to its absence over time. However, the loss of an important species or the loss of multiple species over a short period of time can throw many of these relationships out of balance, possibly bringing about the extinction of other species in the ecosystem.

Organisms also face threats from the physical environment. Such threats range from everyday temperature and moisture conditions to sudden and catastrophic events such as volcanic eruptions and floods. In many ecosystems, seasonal changes bring about life-threatening conditions. In response, some species have developed unique adaptations to overcome acute periods of environmental stress or competition.

During parts of the year when food is scarce or environmental conditions are not conducive to "life as usual," one adaptation certain animals make in response to changes in their habitat is dormancy, which includes hibernation and estivation (also spelled aestivation). Dormancy involves the slowing of life processes that, for animals, results in a lowering of body temperature, reduction of movement, and conservation of energy over a period of time that can range from weeks to months.

There are many causes for dormancy, including changes in environmental temperature, diminished availability of food and water, or reduced quantities of oxygen or carbon dioxide. In the Arctic some animals, polar bears, for example, hibernate during winter. During intense summer heat, some animals in desert areas become dormant to cope with reduced food and water supplies or to preventing overheating.

Many times, the causes of dormancy come in seasonal cycles. Yet, animals that enter dormant states at roughly the same time each year may enter stages of hibernation or estivation even if the environmental conditions that trigger that behaviour no longer exist. For example, Arctic ground squirrels kept in a laboratory still enter a cyclical torpid state, even when food and water are plentiful and temperatures remain constant. Other animals enter dormancy at random times or triggered by other environmental cues, such as lowered concentrations of oxygen and carbon dioxide, that are not seasonally or cyclically induced.

Despite the apparent resilience of life on Earth, one of conservation's most pressing problems is the threat of species extinction. Extinction is defined as the termination of the last member of a species, the death of a species in the wild, or sometimes the decline of a population to the point of being unable to reproduce. Throughout Earth's history, fossil records have revealed periods when more than half of all species on Earth have become extinct within a geologically brief amount of time. Two of the most devastating of Earth's five largest mass extinctions occurred at the Cretaceous–Tertiary (or K-T) boundary some 65.5 million years ago and during the end of the Permian Period, some 260 million to 250 million years ago.

Comparing estimates of the numbers of species becoming extinct during these two events with one another and with extinction rates over the past 500 years, many scientists believe that the planet is currently in the midst of another mass extinction. Environmental scientists concur that the influence of human beings over Earth's ecosystems since the beginning of Industrial Revolution has grown exponentially, and that this is driving Earth's sixth mass extinction, the most rapid of Earth's history. Scientists believe it is the result of human industrialization, which has resulted in enormous increases in greenhouse gas emissions, and the intervention in the world's ecosystems through agricultural expansion, overharvesting of natural resources, and pollution. The world's foremost environmental experts estimate that human activities have increased extinction rates 1,000 times higher than normal "background" extinction rates.

Currently, the primary cause of extinction is the loss of habitat—through encroachment, pollution, overharvesting, overfishing, and the introduction of species into ecosystems that interrupt food webs or predatory chains. Introducing a new species into an environment where it has no natural predators is a dangerous undertaking, since the new species may aggressively feed upon or crowd out native species upon which proper ecosystem functioning depends. Conversely, habitat encroachment by introduced species also destroys species native to the ecosystem that cannot adapt to changes brought on by the introduced species. Both paths often lead to local extinctions and reduced biodiversity. Worldwide, introduced species contribute to the current pattern of accelerated extinction.

Increased extinction rates are bad news for individual species, but they also foreshadow drastic changes to the ecosystems themselves. This is especially true when keystone species (those upon which the existence many other species depend for survival) die out. In tropical forests, fig trees bear fruit year round, providing food for birds and mammals during those times of year when other food sources are unavailable. If the fig trees were to become extinct, numerous animal species might starve and also pass away. This kind of disruption brings about dramatic changes to the structure of an ecosystem. In the case of the fig trees, the loss of this food resource and many of the species that depend on it would simplify food webs, essentially transforming the ecosystem into one with less biodiversity. To prevent the loss of biodiversity, scientists and government officials have begun to intervene in ecosystems to protect species from extinction. In the United States, the Endangered Species Act (ESA) provides protection for thousands of threatened or endangered species and the habitats they live in. Invasive species have often been forcibly removed from an ecosystem to strengthen threatened species, and habitats have been protected from human encroachment to prevent the loss of essential plants and animals necessary for the survival of the endangered species. The practice of habitat restoration has enabled many individuals belonging to species to survive until adulthood, produce more offspring, and thus increase the size of their populations. International treaties such as the International Convention on Trade in Endangered Species of Fauna and Flora, the International Convention on the Regulation of Whales, and the Convention for the Protection and Development of the Marine Environment of the Wider Caribbean Region are just a few of the efforts taken by the countries of the world to address this issue.

The study of ecology and the activities of the conservation movement depend on the work of numerous key individuals. These ecologists and environmentalists have come from diverse educational backgrounds. Pioneers in the fields of conservation and ecology include John Muir (1838–1914), a naturalist and promoter of forest conservation. Muir was a key force in the creation of Sequoia and Yosemite National Parks in California. Another conservationist, Gifford Pinchot (1865–1946) was deeply involved in forestry and conservation as a member of the

National Forest Commission of the National Academy of Sciences, and as chief of the newly created Forest Service in the Department of Agriculture under three presidents. Rachel Carson (1907–1964), author of *The Sea Around Us* and *Silent Spring*, captured the attention of the world with her impassioned pleas for action on environmental pollution.

Other early voices for conservation were Theodore Jesse Hoover and Aldo Leopold. Hoover (1871–1955), an American mining engineer, naturalist, and educator, made numerous contributions to conservation of natural resources, and authored several books and numerous articles on his work in mining and metallurgy. Aldo Leopold (1887–1948), an American environmentalist, worked for the U.S. Forest Service in the Southwestern United States. He was instrumental in the creation of the Gila Wilderness Area in New Mexico, a director of the Audubon Society, and a founder of the Wilderness Society.

More recently, United States biologist and educator Barry Commoner (1917–) issued warnings in the 1950s about threats to the environment from nuclear weapons, pesticides, toxic chemicals, and inadequate waste management. Paul R. Ehrlich (1932–), also from the United States, warned of the dangers of rampant overpopulation. His controversial 1968 book, *The Population Bomb*, introduced a whole generation to the science of ecology.

Al Gore (1948–), the 45th Vice President of the United States, has spent years attempting to educate the world about the dangers of global warming. Gore, along with the International Panel for Climate Change, was awarded the Nobel Prize for Peace in 2007 for his tireless efforts communicating the notion that global warming threatens Earth's current future generations.

Jane Lubchenco (1947–), an American environmental scientist and marine ecologist, became the first woman to serve as the administrator of the National Ocean and Atmospheric Administration (NOAA) and as the undersecretary for oceans and atmosphere after a distinctive career devoted to environmental conservation and preservation. Raman Sukumar (1955–) is a noted Indian ecologist, renowned for his work with Asian elephants to prevent the extinction of the species, as well as his research on climate change, tropical forests, and wildlife conservation.

These extraordinary people—and many others—have been committed to the understanding of ecology and to the conservation of Earth's biodiversity. Through education and direct action, each one has performed the essential work of protecting species in crisis.

CHAPTER 1

THE PATHOLOGY OF EXTINCTION

Conservation is the study of the loss of Earth's biological diversity and the ways this loss can be prevented. Biological diversity, or biodiversity, is the variety of life either in a particular place or on the entire Earth, including its ecosystems, species, populations, and genes. Conservation thus seeks to protect life's variety at all levels of biological organization.

Species extinction is the most obvious aspect of the loss of biodiversity. For example, species form the bulk of the examples in a comprehensive assessment of the state of the planet published in the early 21st century by the Millennium Ecosystem Assessment, an international effort coordinated by the United Nations Environment Programme. The subject of conservation is broader than this, however. Even a species that survives extinction can lose much of its genetic diversity as local, genetically distinct populations are lost from most of the species' original range. Furthermore, ecosystems may shrink dramatically in area and lose many of their functions, even if their constituent species manage to survive. Conservation is involved with studying all these kinds of losses, understanding the factors responsible for them, developing techniques to prevent losses, and, whenever possible, restoring biodiversity.

Conservation is a crisis discipline, one demanded by the unusual rates of loss. It is also a mission-driven discipline. By analogy, ecology and conservation have the same relationship as physiology and medicine. Human physiology studies the workings of the human body, whereas medicine is mission-oriented and aims to understand what goes wrong and how to treat it.

Conservation is often considered a purely biological topic, as exemplified by major scientific journals with titles such as *Conservation Biology* and *Animal Conservation* as well as college textbooks with such titles as *Principles of Conservation Biology* and *Essentials of Conservation Biology*. However, because the underlying cause of the loss of biodiversity is increasing human activity, conservation must inevitably involve human interactions. Many of the techniques to prevent the loss of biodiversity involve issues of economics, law, social sciences, and religion—all of which are covered by the journals and textbooks previously cited.

According to the best estimates of the world's environmental experts, human activities have driven species to extinction at rates perhaps 1,000 times the natural, or background, rate, and future rates of extinction will likely be higher. To show how the experts arrived at these conclusions, it is necessary to pose and attempt to answer a series of extremely difficult questions. How many species are there? How fast were species disappearing before human activity became pervasive? How fast are they becoming extinct at present? And finally, it is necessary to ask a further question: What does the future hold for extinctions if current trends continue?

HOW MANY SPECIES ARE THERE?

Any absolute estimate of extinction rate, such as extinctions per year, requires knowledge of how many species there are. Unfortunately, this number is not known with any great degree of certainty, and the problems of estimating it are formidable. Taxonomists have described—that is, have given names to—about 1.5 million species. Only about 100,000 of them, comprising terrestrial vertebrates, some flowering plants, and attractive and collectible invertebrates such as butterflies and snails, are popular enough

Club fungus (Claveria) *growing in soil.* Ken Brate/Photo Researchers

for taxonomists to know well. Birds are exceptionally well-known—there are roughly 10,000 bird species, with only one or two new species being added each year.

Those who describe species cannot always be certain that the specimen in hand has not been given a name by someone else in a different country and sometimes even in a different century. Consequently, some taxonomic groups may have more names assigned to them than constituent species, which would result in erroneously high species estimates. Potentially much more serious as a source of error is the fact that some species groups have relatively few named members compared with the numbers that experts think exist in those groups. For example, taxonomists have only sparsely sampled some potentially rich communities, such as the bottom of the deep ocean and the canopies of rainforests.

One estimate of how many species might still be undescribed involves a comparison of fungi and flowering plants (angiosperms). In Great Britain, where both groups are well-known, there are six times as many named species of fungi as of flowering plants. If this ratio applies worldwide, the world total of about 300,000 species of flowering plants, which are fairly well-known globally, predicts a total of about 1.8 million species of fungi, which are not. Because only about 70,000 species of fungi currently have names, the prediction would suggest that these named species constitute only $^{1}/_{25}$ of the species that exist. Conversely, samples from poorly known parts of the world suggest that the named species constitute a much larger fraction. Although this estimate of the number of species of fungi is almost certainly too high,

Dandelion (Taraxacum officinale). Alan Punton/A to Z Botanical Collection

it nonetheless suggests that there are large numbers of unknown species.

For insects, there are about 1 million described species, yet estimates of how many insect species exist range from 10 to 100 times this number. As an example of how part of one such estimate was made, an entomologist collected a large sample of canopy-dwelling beetles from one species of tropical tree. The tree species had 163 beetle species specific to it. There are about 50,000 tree species in the world's tropical forests, so simple multiplication predicts roughly 8 million species of tropical forest-canopy beetles. Because 40 percent of described insects are beetles, the total number of tropical forest-canopy insects could be 20 million. Adding half that number for insect species found on the ground beneath the trees gives an estimated grand total of 30 million species of tropical forest insects alone.

This example is critically dependent on the first number—the 163 species found only on the one species of tree. In addition, other calculations suggest that this estimate is too high. The number of insect species in tropical forests is more likely to be between 7 and 15 million, which still means that most of Earth's species would be tropical forest insects. Given the considerable uncertainties, many scientists consider that the total of all species is very roughly 10 million, a number that likely translates to "somewhere between 5 and 20 million with arguments for numbers both lower and higher."

An obvious concern follows regarding the usefulness of such calculations as a basis for assessing the loss of species. Any absolute estimate of species extinctions must be extrapolated from the 100,000 well-known species of living plants and animals, to the roughly 1.5 million described species, to the likely grand total of very roughly 10 million.

Because of uncertainties about the total number of living species, published statements regarding the total number of species that become extinct per year or per day can vary a hundredfold.

Another approach to assessing species loss is to derive relative estimates—estimates of the proportion of well-known species that become extinct in a given interval. Estimating such proportions is the basis for the remainder of the discussion on rates of extinction, but it raises a critical concern of its own: are these proportions actually typical of most species that are still undescribed? They are likely to be so if extinction rates in widely different species groups and regions turn out to be broadly similar.

There also is another way in which estimates of extinctions can be made relative. Extinctions have always been a part of Earth's history. It is possible to make any estimates of massive future extinction relative to that history.

CALCULATING BACKGROUND EXTINCTION RATES

To discern the effect of modern human activity on the loss of species requires determining how fast species disappeared in the absence of that activity. Studies of marine fossils show that species last about 1–10 million years. Assume that all these extinctions happened independently and gradually (i.e., the "normal" way) rather than catastrophically, as they did at the end of the Cretaceous Period about 65 million years ago, when dinosaurs and many other land and marine animal species disappeared. On that basis, if one followed the fates of one million species, one would expect to observe about 0.1–1 extinction per year—in other words, one species going extinct every 1–10 years.

Human life spans provide a useful analogy to the foregoing. If humans live for about 80 years on average, all things being equal one would expect that 1 in 80 individuals should die each year under normal circumstances. If, however, many more than 1 in 80 were dying each year, something would be abnormal. There might be an epidemic, for instance.

To make comparisons of present-day extinction rates conservative, assume that the normal rate is just one extinction per million species per year. This then is the benchmark—the background rate against which one can compare modern rates. For example, given a sample of 10,000 living described species (roughly the number of modern bird species), one should see one extinction every 100 years. Comparing this to the actual number of extinctions within the past century provides a measure of relative extinction rates.

The estimates of the background extinction rate described above derive from the abundant and widespread species that dominate the fossil record. By contrast, the species most likely to become extinct today are rare and local. Thus, the fossil data might underestimate background extinction rates. Importantly, however, these estimates can be supplemented from knowledge of speciation rates—the rates that new species come into being—of those species that often are rare and local. These rates cannot be much less than the extinction rates, or there would be no species left.

To explore the idea of speciation rates, one can refer again to the analogy of human life spans and ask: How old are my living siblings? The answer might be anything from that of a newborn to that of a retiree living out his or her last days. The average age will be midway between them—that is, about half a lifetime. Ask the same question for a mouse, and the answer will be a few months; of long-living

Representative apes (superfamily Hominoidea). Encyclopædia Britannica, Inc.

trees such as redwoods, perhaps a millennium or more. The age of one's siblings is a clue to how long one will live.

Species have the equivalent of siblings. They are the species' closest living relatives in the evolutionary tree—something that can be determined by differences in the DNA. The closest relative of human beings is the bonobo (*Pan paniscus*), whereas the closest relative of the bonobo is the chimpanzee (*P. troglodytes*). Taxonomists call such related species sister taxa, following the analogy that they are splits from their "parent" species.

The greater the differences between the DNA of two living species, the more ancient the split from their common ancestor. Studies show that these accumulated differences result from changes whose rates are, in a certain fashion, fairly constant—hence, the concept of the molecular clock—which allows scientists to estimate the time of the split from knowledge of the DNA differences. For example, from a comparison of their DNA, the bonobo and the chimpanzee appear to have split one million years ago, and humans split from the line containing the bonobo and chimpanzee about six million years ago.

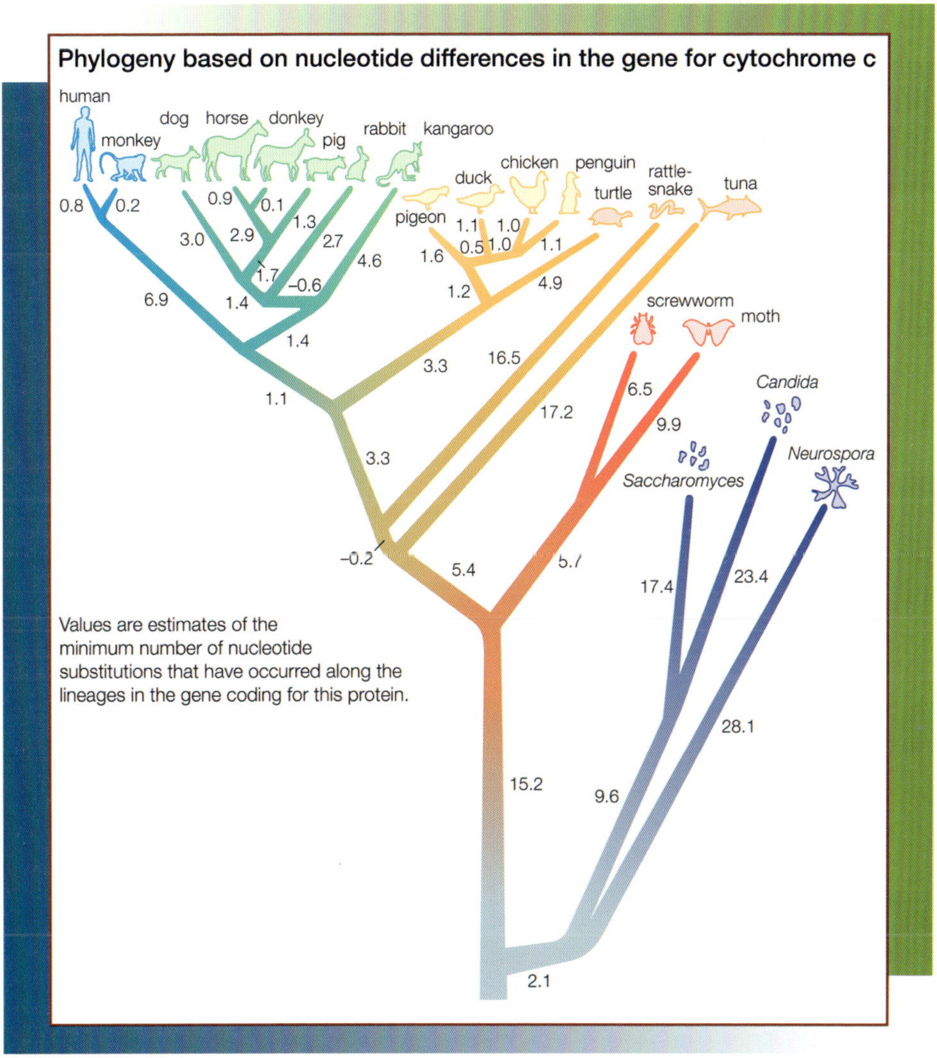

Encyclopædia Britannica, Inc.

The advantage of using the molecular clock to determine speciation rates is that it works well for all species, whether common or rare. It works for birds and, in the previous example, for forest-living apes, for which only scant fossils have been recovered. In the preceding example, the bonobo and chimpanzee split a million years ago, suggesting such species' life spans are, like those of the abundant and widespread marine species discussed earlier, on million-year timescales, at least in the absence of modern human actions that threaten them. This is just one example, however. Is there evidence that speciation can be much more rapid?

Until recently, there seemed to be an obvious example of a high rate of speciation—a "baby boom" of bird species. Its existence allowed for the possibility that the high

Western bluebird (Sialia mexicana). Herbert Clarke

rates of bird extinction that are observed today might be just a natural pruning of this evolutionary exuberance.

On either side of North America's Great Plains are 35 pairs of sister taxa including western and eastern bluebirds (*Sialia mexicana* and *S. sialis*), red-shafted and yellow-shafted flickers (both considered subspecies of *Colaptes auratus*), and ruby-throated and black-chinned hummingbirds (*Archilochus colubris* and *A. alexandri*). According to the rapid-speciation interpretation, a single mechanism seemed to have created them all. Each pair of sister taxa had one parent species ranging across the continent. Then a major advance in glaciation during the latter part of the Pleistocene Epoch (2.6 million to 11,700 years ago) split each population of parent species into two groups. Each pair of isolated groups evolved to become two sister taxa, one in the west and the other in the east. Finally, the ice retreated, and, as the continent became warm enough, about 10,000 years ago, the sister taxa expanded their ranges and, in some cases, met once again.

The story, while compelling, is now known to be wrong. Molecular data show that, on average, the sister taxa split 2.45 million years ago. This means that the average species life span for these taxa is not only very much older than the rapid-speciation explanation for them requires but is also considerably older than the one-million-year estimate for the extinction rate previously suggested as a conservative benchmark.

Molecular-based studies find that many sister species were created a few million years ago, which suggests that species should last "a few million" years, too. Indeed, they suggest that the background rate of one extinction among a million species per year may be too high. Nevertheless, this rate remains a convenient benchmark against which to compare modern extinctions.

RECENT EXTINCTION RATES

To what extent has modern human activity increased extinction rates above the background rate? This discussion presents five well-known case histories of recent extinctions. From them, some general features can be deduced about recent extinctions that also provide clues to the future.

Pacific Island Birds

Polynesians reached such remote Pacific islands as the Hawaiian Islands, New Zealand, and Easter Island—Earth's last habitable areas for settlement—within the past 2,000 years. Over that period they left unambiguous evidence that their activity caused many species of birds to become extinct. The bones of many species persist into, but not through, archaeological layers that also contain evidence of human presence. No species is known to have disappeared in the longer intervals before first contact. The Polynesian settlers likely ate the large, probably unwary, and often flightless species. They also introduced pigs and rats to islands far too remote to have acquired native land mammals. Native birds, their eggs, and their young would have been easy pickings for the rats, and the pigs would have destroyed the ground cover of the forests. With only Stone Age technology, the settlers may have exterminated as many as 2,000 bird species, some 17 percent of the world total. Locally, they often exterminated all the bird species they encountered.

In the Hawaiian Islands, for example, scientists have described 43 species only from their bones, a number that has increased as new extinct species have been discovered. Because bird bones are fragile and easily destroyed, all the extinct species may never be found. Nevertheless, the number that remain unknown can be calculated.

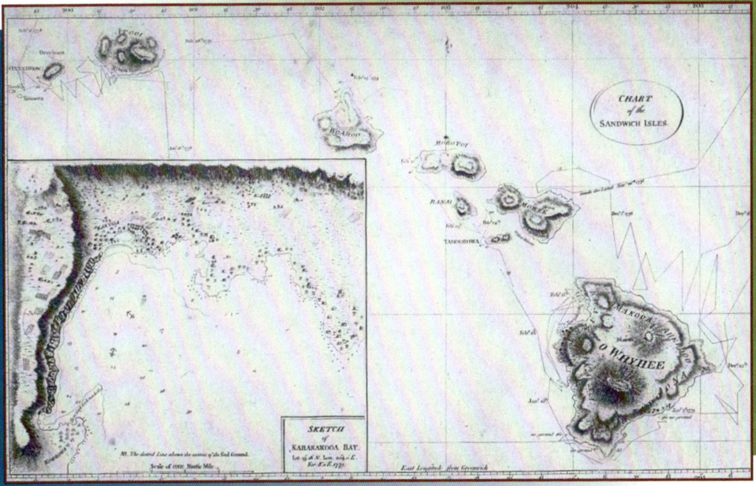

Sandwich, or Hawaiian, Islands chart produced by James Cook's third Pacific Ocean voyage in 1779. Cook arrived on islands teeming with life. Introducing destructive new species and depleting native plants and animals, however, irrevocably upset the islands' ecological balance. Time & Life Pictures/Getty Images

The British explorer James Cook found the Hawaiian Islands and their Polynesian settlers in 1778. With the peace that followed Great Britain's defeat of France in 1815, Cook's descriptions of whaling opportunities in the region led to increasing contact with Europe and North America. New colonists not only depleted the whales but also introduced cattle and goats to the islands for food. Like pigs, these alien herbivores destroyed native plants and greatly reduced natural habitats. Naturalists of the time described 18 bird species that did not survive this onslaught, so the total count of extinctions rises to 101. This still is an underestimate, because the 19th-century naturalists missed some species. On Molokai, for example, they recorded hearing a rail, but there is no specimen of it.

On the Hawaiian Islands today there remain a dozen species that are so extremely rare that the chances of saving them are poor. Some may already be extinct. Another dozen species are rare and have uncertain futures. Of the estimated 136 bird species that lived on the islands before human settlement, only 11 are common enough to suggest that their future is secure.

If such estimates are expanded across the eastern Pacific, it will appear that the Polynesians exterminated 500–1,000 species of birds. But it is also possible to take a different approach, one that considers what species should occur on every island in the group. Rails, for example, seem most adept at reaching even such remote islands as Henderson and Lisianski. Where scientists have searched for bird bones, they have found fossil rails. It seems likely that every sufficiently large island probably housed a unique species of rail. Most of these islands, but not the Hawaiian Islands, which were likely too remote, also housed unique species of doves and parrots, most of which are now extinct. Even before small-bodied species are counted, such studies suggest that Polynesians exterminated at least 1,000 species of birds. Some scientists put the number at 2,000 or even more.

The Polynesian and European colonizations of the Pacific exterminated species other than birds, of course. The Hawaiian Islands have 980 native plants. Of these, 84 are now extinct, and another 133 have wild populations that number fewer than 100 individuals each. It may never be known how many other species were exterminated as the Polynesians cleared land for their crops and likely burned dry forests.

In the 20th century, misguided individuals on a number of the Pacific islands introduced an African land snail, *Achatina fulica*, for food. It became a pest. So, like the song about the old woman who swallowed a fly, and then a spider to catch it, and so forth, a predatory snail, *Euglandina rosea*, was released to control

Dodo (Raphus cucullatus). Encyclopædia Britannica, Inc.

the *Achatina*. The predatory snail preferred native *Achatinella* and *Partula* snails instead, driving many species to extinction.

There is nothing special about the islands of the Pacific that makes their animals and plants prone to extinction. The most famous recent extinction of all, the extermination of the dodo (*Raphus cucullatus*) by humans and their introduced animals, was one of 33 species of birds, 30 species of land snails, and 11 species of reptiles lost from Mauritius, Rodrigues, and Réunion in the Indian Ocean in the past three centuries. Over the same period, St. Helena and Madeira in the Atlantic Ocean lost 36 species of land snails as a result of human activity. These island extinctions raise an obvious question: Are extinctions predominantly on islands and not continents? Although the disappearance of island species does demonstrate a pattern of sensitivity to becoming extinct, the next example shows the answer to the preceding question to be "no."

Flowering Plants

A small area at the southern tip of Africa houses an extraordinarily rich and unusual variety of plants. It has several types of vegetation, of which the fynbos (a species-rich scrubland of southwestern South Africa) contributes the most species. Of roughly 8,500 species, 36 have become extinct in the past century, and another 618 species are threatened. (In the field of conservation, the term *threatened* has a specific, technical meaning. It comes from the World Conservation Union's [IUCN's] Red Lists, the lists of species that are at risk of extinction. A species listed as "threatened" has a high probability of extinction in the wild within the next few decades if nothing is done to prevent it.) For the fynbos, invading alien plants—particularly various species of Australian wattle trees (acacia)—and the conversion of natural areas to agriculture are the two major causes of species extinction and endangerment.

Freshwater Mussels and Clams

Terrestrial ecosystems are far from being the only places where recent extinctions have occurred. The Mississippi and St. Lawrence river basins were home to 297 North American species of the bivalve mollusk families Unionidae and Margaritiferidae. Of these, 21 have become extinct in the past century, and another 120 species are in danger of extinction. During this same period, engineers have extensively dammed and channeled North America's rivers. The Tennessee River, for example, is dammed along its entire length from Knoxville, Tenn., until it joins the Ohio River. Changing flowing rivers to effectively static reservoirs completely changes the habitats of the species that live in them. In addition, poor farming practices have greatly increased nutrient and pesticide runoff from croplands, and sewage effluent and runoff from cities have exacerbated the problem. Freshwater bivalves are sessile filter feeders and very sensitive to changes in water quality.

Many of these bivalve species in their juvenile stages parasitize freshwater fish and so are dependent on their survival as well. Some freshwater bivalves are harvested for the cultured pearl trade. And, as if these were not sufficient threats, two introduced species, the Asian clam (*Corbicula fluminea*) and the zebra mussel (*Dreissena polymorpha*), have rapidly expanded their North American ranges, outcompeting native species.

Freshwater Fish

Some of the changes to North America's rivers that threaten their native bivalves have also seriously harmed the continent's freshwater fish. Of the approximately 950 species of freshwater fish of the United States, Canada, and Mexico, about 40 have become extinct in the past century. They have disappeared from a wide range of habitats—northern lakes, southern streams, wetlands, and particularly springs. Springs are often small, isolated habitats, which favour the evolution of unique species, but those features make the habitats extremely vulnerable to human disturbance. Introduced species—from the stocking of fish and frogs for food and of fish for sport and from accidental aquarium releases—are also factors in 18 of the 40 extinctions. The introduced species can outcompete native fish, sometimes feed on them, and in some cases hybridize with them. Hybridization may seem a surprising cause of extinction. Populations that have evolved in isolation, however, may no longer retain their morphologic and genetic distinctiveness when mixed with similar species from elsewhere.

The southeastern United States holds a little over half of North America's fish, many of which have small geographic ranges, living in one or just a few rivers. Extensive damming and pollution of southern Appalachian mountain streams have exterminated 4 species and threatened another 80.

Australian Mammals

Compared with the mammals found on other continents, Australian mammals have proved especially vulnerable to extinction. At least 60 mammals have become extinct worldwide in the past two centuries, about one-third of them from Caribbean islands. Of the remainder, 18 mammals were native to Australia, where they constituted about 6 percent of the terrestrial animal species prior to the British colonization of the continent beginning in the late 18th century. (Aboriginal peoples reached Australia at least 42,000 years earlier and, like the Polynesians across the eastern Pacific, eliminated many native species.) More than 40 other living mammal species in Australia have lost at least half their former ranges, and some survive only on protected offshore islands.

Some extinctions have taken place in what is now the wheat belt of the southern tip of Western Australia, which retains only 5 percent of its natural woodland vegetation cover. Others have occurred in the southern

The Tasmanian tiger—also called the thylacine or Tasmanian wolf—was a carnivorous marsupial that became extinct in the 1930s. It was once found in Australia and on the island of New Guinea. Encyclopædia Britannica, Inc.

arid zone, an area of mostly desert dominated by spinifex grass—and very few people. Nonetheless, this is an area where domestic grazing animals have destroyed the natural vegetation and caused extensive soil erosion. Moreover, European rabbits (*Oryctolagus cuniculus*) introduced in the mid-19th century are competitors of the native mammals, and the red fox (*Vulpes vulpes*) introduced about the same time has likely destroyed native small-mammal populations even in remote areas. Where foxes are absent in Australia or where their numbers have been controlled, the native mammals seem to have fared better. In addition, farmers and graziers have greatly altered the natural fire regimes—the roles that fire plays in ecosystems in the absence of human intervention—with consequent changes to the vegetation and the animals that depend on it.

CALCULATING THE RELATIVE RATES OF EXTINCTION

To show how extinction rates are calculated, the discussion will focus on the group that is taxonomically the best-known—birds. The modern process of describing bird species dates from the work of the 18th-century Swedish botanist Carolus Linnaeus in 1758. Even at that time, two of the species that he described were extinct, including the dodo. The 1800s was the century of bird description—7,079 species, or roughly 70 percent of the modern total, were named. Of those species, 39 became extinct in the subsequent 100 years. The corresponding extinction rate is 55 extinctions per million species per year. More than 100 of those 7,079 species are classified as critically endangered—the most threatened category of species listed by the IUCN—or else are dependent on conservation efforts to protect them. When similar calculations are done on bird species described in other centuries, the results are broadly similar.

The same approach can be used to estimate recent extinction rates for various other groups of plants and animals. The calculated extinction rates, which range from 20 to 200 extinctions per million species per year, are high compared with the benchmark background rate of 1 extinction per million species per year, and they are typical of both continents and islands, of both arid lands and rivers, and of both animals and plants.

Although less is known about invertebrates than other species groups, it is clear from the case histories discussed earlier that high rates of extinction characterize both the bivalves of continental rivers and the land snails on islands. In fact, there is nothing special about the life histories of any of the species in the case histories that make them especially vulnerable to extinction. Indeed, what is striking is how diverse they are. The same is true for where the species live—high rates of extinction occur in a wide range of different ecosystems.

To draw reliable inferences from these case histories about extinctions in other groups of species requires that these be representative and not selected with a bias toward high extinction rates. In reviewing the list of case histories, it seems hard to imagine a more representative selection of samples. With high statistical confidence, they are typical of the many groups of plants and animals about which too little is known to document their extinction. In short, one can be certain that the present rates of extinction are generally pathologically high even if most of the perhaps 10 million living species have not been described or if not much is known about the 1.5 million species that have been described.

ENDANGERED SPECIES

Any species of plant, animal, or other organism that is at risk of extinction because of a sudden rapid decrease in its population or a loss of its critical habitat is referred to as an endangered species. Previously, any species of plant or animal that was threatened with extinction could be called an endangered species. The need for separate definitions of "endangered" and "threatened" species has resulted in the development of various categorization systems, each containing definitions and criteria by which a species can be classified according to its risk of extinction. As a rule, a range of criteria must be analyzed before a species can be placed in one category or another. Often such categorization systems are linked directly to national legislation, such as the United States Endangered Species Act (ESA) or the Canadian Species at Risk Act (SARA). In addition, regional agreements—such as the European Union's Habitats Directive (Council Directive 92/43/EEC)—or international conservation agreements—such as the Convention on the Conservation of Migratory Species of Wild Animals (CMS) or the Convention on

The polar bear (Ursus maritimus) *is one of several prevalent endangered species protected by the Endangered Species Act.* Shutterstock.com

International Trade in Endangered Species of Wild Fauna and Flora (CITES)—are connected to species assessment systems. One of the most recognized independent international systems of species assessment is the Red List of Threatened Species, created by the International Union for Conservation of Nature (IUCN).

Roughly 99 percent of threatened species are at risk because of human activities alone. By the early 21st century, it could be said that human beings (*Homo sapiens*) are the greatest threat to biodiversity. The principal threats to species in the wild are:

1. Habitat loss and habitat degradation

2. The spread of introduced species (that is, non-native species that negatively affect the ecosystems they become part of)

3. The growing influence of global warming and chemical pollution

4. Unsustainable hunting

5. Disease

Although some of these hazards occur naturally, most are caused by human beings and their economic and cultural activities. The most pervasive of these threats, however, is habitat loss and degradation—that is, the large-scale conversion of land in previously undisturbed areas driven by the growing demand for commercial agriculture, timber extraction, and infrastructure development. With the rates of loss being highest in some of the most biologically diverse regions on Earth, there is a perpetual battle to manage destructive activities while limiting the effect such restrictions may have on the well-being of local communities. The relative importance of each threat differs among and between taxa. So far, although incidental mortality, temporary or limited human disturbance, and persecution have caused limited reductions in the total number of species, these phenomena can be serious for some susceptible groups. In addition, global warming has emerged as a widespread threat, and much research is being

Amphibians such as this Albino axolotl (Ambystoma mexicanum) *are especially sensitive to changes in the environment.* Jane Burton/Bruce Coleman Ltd.

conducted to identify its potential effects on specific species, populations, and ecosystems.

Against this backdrop of threats related to urban expansion and food production, the unsustainable harvest of animal and plant products for traditional medicine and the pet trade is a growing concern in many parts of the world. These activities have implications on local ecosystems and habitats by exacerbating population declines through overharvesting. In addition, they have cross-border repercussions in terms of trade and illegal trafficking.

PREDICTING THE FUTURE RATES OF EXTINCTION

Not only do the five case histories demonstrate recent rates of extinction that are tens to hundreds of times higher than the natural rate, but they also portend even higher rates for the future. For every recently extinct species in a major group, there are many more presently threatened species. IUCN Red Lists in the early years of the 21st century reported that about 12 percent (i.e., about 1,200 species) of the roughly 10,000 living bird species are at risk of extinction. Furthermore, information in the same source indicates that this percentage is lower than that for mammals, reptiles, fish, flowering plants, or amphibians. For example, 16 percent of flowering plants are deemed threatened among the roughly 300,000 described species. Plant conservationists estimate that 100,000 plant species remain to be described, the majority of which will likely turn out to be rare and very local in their distribution. The latter characteristics explain why these species have not yet been found and make the species particularly vulnerable to extinction. Should any of these plants be described, they are likely to be classified as threatened, so the figure of 16 percent is an underestimate.

The Nature Conservancy in 1997 issued a report card on species for the United States that covered 13 groups of animals and plants. Using standards that were different, though broadly comparable, to those of the IUCN, the Nature Conservancy found that nearly 15 percent of birds were in danger of extinction (again the smallest percentage); between 16 and 18 percent of mammals, butterflies, reptiles, and dragonflies; up to about 40 percent of freshwater fish and amphibians; and more than two-thirds of freshwater mussels.

Giant panda (Ailuropoda melanoleuca) *feeding in a bamboo forest, Sichuan (Szechwan) province, China. The IUCN Red List of Threatened Species reports that fewer than 2,500 mature pandas may remain in the wild.* Wolfshead—Ben Osborne/Ardea London

Once again choosing birds as a starting point, let us assume that the threatened species might last a century—this is no more than a rough guess. The 1,200 species of birds at risk would then suggest a rate of 12 extinctions per year on average for the next 100 years. That translates to 1,200 extinctions per million species per year, or 1,200 times the benchmark rate. This number, uncertain as it is, suggests a massive increase in the extinction rate of birds and, by analogy, of all other species, because the percentage of species at risk in the bird group is estimated to be lower than the percentages in other groups of animals and plants. Improving on this "rough guess" requires a more detailed assessment of the fates of different sets of species.

IUCN RED LIST OF THREATENED SPECIES

One of the most well-known objective assessment systems for declining species is the approach unveiled by the International Union for Conservation of Nature (IUCN) in 1994. It contains explicit criteria and categories to classify the conservation status of individual species on the basis of their probability of extinction. This classification is based on thorough, science-based species assessments and is published as the IUCN Red List of Threatened Species, more commonly known as the IUCN Red List. It is important to note that the IUCN cites exacting criteria for each of these categories, and the descriptions given in the following text have been condensed to highlight two or three of the category's most salient points. In addition, three of the categories (CR, EN, and VU) are contained within the broader notion of "threatened." The list recognizes nine categories of species status:

1. Extinct (EX), a designation applied to species in which the last individual has died or where systematic and time-appropriate surveys have been unable to log even a single individual

2. Extinct in the Wild (EW), a category containing those species whose members survive only in captivity or as artificially supported populations far outside their historical geographic range

3. Critically Endangered (CR), a category containing those species that possess an extremely high risk of extinction as a result of rapid population declines of 80 to more than 90 percent over the previous 10 years (or three generations), a current population size of fewer than 50 individuals, or other factors

4. Endangered (EN), a designation applied to species that possess a very high risk of extinction as a result of rapid population declines of 50 to more than 70 percent over the previous 10 years (or three generations), a current population size of fewer than 250 individuals, or other factors

5. Vulnerable (VU), a category containing those species that possess a very high risk of extinction as a result of rapid population declines of 30 to more than 50 percent over the previous 10 years

(or three generations), a current population size of fewer than 1,000 individuals, or other factors

6. Near Threatened (NT), a designation applied to species that are close to becoming threatened or may meet the criteria for threatened status in the near future

7. Least Concern (LC), a category containing species that are pervasive and abundant after careful assessment

8. Data Deficient (DD), a condition applied to species in which the amount of available data related to its risk of extinction is lacking in some way. Consequently, a complete assessment cannot be performed. Thus, unlike the other categories in this list, this category does not describe the conservation status of a species.

9. Not Evaluated (NE), a category used to include any of the nearly 1.6 million species described by science but not yet assessed by the IUCN.

The IUCN system uses five quantitative criteria to assess the extinction risk of a given species. In general, these criteria consider:

1. How quickly the population is declining

2. The geographic range of the species

3. Whether the species already possesses a small population size

4. Whether the species is very small or lives in a restricted area

5. Whether the results of a quantitative analysis indicates a high probability of extinction in the wild

All else being equal, a species experiencing a 90 percent decline over 10 years (or three generations), for example, would be classified as critically endangered. Likewise, another species undergoing a 50 percent decline over the same period would be classified as endangered, and one experiencing a 30 percent reduction over the same time frame would be considered vulnerable. It is important to understand, however, that a species cannot be classified by using one criterion alone. It is essential for the scientist doing the assessment to consider all five criteria to determine the status. Each year, thousands of scientists around the world assess or reassess species according to these criteria, and the IUCN Red List is subsequently updated with these new data once the assessments have been checked for accuracy to help provide a continual spotlight on the status of the world's species.

The IUCN Red List brings into focus the ongoing decline of Earth's biodiversity and the influence humans have on life on the planet. It provides a globally accepted standard with which to measure the conservation status of species over time. By 2008, 44,838 species had been assessed by using the IUCN Red List categories and criteria. Of these, 16,928 species of plants, animals, and others fell into the threatened categories (CR, EN, and VU), with 7,744 species considered either endangered or critically endangered. Today the list itself is an online database available to the public. Scientists can analyze the percentage of species in a given category and the way these percentages change over time; they can also analyze the threats and conservation measures that underpin the observed trends.

Inexorable Species Declines

Some threatened species are declining rapidly. For a proportion of these, eventual extinction in the wild may be so certain that conservationists may attempt to take them into captivity to breed them. Recent examples include the California condor (*Gymnogyps californianus*), which has been reintroduced into the wild with some success, and the alala (or Hawaiian crow, *Corvus hawaiiensis*), which has not. Other species have not been as lucky. In the early 21st century an exhaustive search for the baiji (*Lipotes vexillifer*), a species of river dolphin found in the Yangtze River, failed to find any. The dolphin had declined in numbers for decades, and efforts to keep the species alive in captivity were unsuccessful.

Some species have no chance for survival even though their habitat is not continuously declining. Those who claim that extraordinary species such as the famous Loch Ness monster ("Nessie") have long been surviving as solitary individuals or very small mating populations overlook the basics of sexual reproduction. If a species, be it proved or only rumoured to exist, is down to one individual—as some rare species are—it has no chance. The odds are not much better if there are a few more individuals. If one breeding pair exists and if that pair produces two young—enough to replace the adult numbers in the next generation—there is a 50-50 chance that those young will be both male or both female, whereupon the population will go extinct. Even if they were male and female, they would be brother and sister, and their progeny would likely suffer from a variety of genetic defects. In the case of two breeding pairs—and four young—the chance is one in eight that the young will all be of the same sex. Although these are better odds, if the species plays this game every generation, only replacing its numbers, over many generations the probability is high that one generation will have four young of the same sex and so bring the species to extinction. Thus, for just one Nessie to be alive today, its numbers very likely would have to have been substantial just a few decades ago. These and related probabilities can be explored mathematically, and such models of small populations provide crucial advice to those who manage threatened species.

Field studies of very small populations have been conducted. For example, small islands off the coast of Great Britain have provided a half-century record of many bird species that traveled there and remained to breed. This record shows that most small populations formed by individuals that colonized from the mainland persisted for a few years to decades before going extinct.

Surviving but Threatened Small Populations

A broad range of environmental vagaries, such as cold winters, droughts, disease, and food shortages, cause population sizes to fluctuate considerably from year to year. The behaviour of butterfly populations is well studied in this regard. Because their numbers can decline from one year to the next by 99 percent, even quite large populations may be at risk of extinction. In the case of smaller populations, the Nature Conservancy reported that, of about 600 butterfly species in the United States, 16 species number fewer than 3,000 individuals and another 74 species fewer than 10,000 individuals. Butterfly numbers are hard to estimate, in part because they do fluctuate so much from one year to the next, but it is clear that such natural fluctuations could reduce low-population species to numbers that would make recovery unlikely.

The role of population fluctuations has been dissected in some detail in a long-term study of the Bay checkerspot butterfly (*Euphydryas editha bayensis*) in the grasslands above Stanford University in Palo Alto, Calif. In 1960 scientists began following the fate of several local populations of the butterfly at a time when grasslands around San Francisco Bay were being lost to housing developments. The populations were themselves isolated from each other, with only little migration between them. In succeeding decades small populations went extinct from time to time, but immigrants from two larger populations reestablished them. Nonetheless, in 1991 and 1998 first one and then the other larger population became extinct. At their peaks the former had reached almost 10,000 individuals and the latter about 2,000 individuals, although this second population was less variable from year to year. The Bay checkerspot still lives in other places, but the study demonstrates that relatively small populations of butterflies (and, by extension, other insects) whose numbers undergo great annual fluctuations can become extinct quickly.

What are the consequences of these fluctuations for future extinctions worldwide? Simply put, habitat destruction has reduced the majority of species everywhere on Earth to smaller ranges than they enjoyed historically. Where these ranges have shrunk to tiny protected areas, species with small populations have no possibility of expanding their numbers significantly, and quite natural fluctuations (along with the reproductive handicaps of small populations) can exterminate species.

In sum, most of the presently threatened species will likely not survive the 21st century. They may already be declining inexorably to extinction. Alternately, their populations may number so few that they cannot survive more than a few generations or may not be large enough to provide a hedge against the risk that natural fluctuations will eventually lead to their extinction.

Because some threatened species will survive through good luck and others by good management of them, estimates of future extinction rates that do not account for these factors will be too high. A factor having the potential to create more serious error in the estimates, however, consists of those species that are not now believed to be threatened but that could become extinct. For example, the 2006 IUCN Red List for birds added many species of seabirds that formerly had been considered too abundant to be at any risk. Over the previous decade or so, the growth of longline fishing, a commercial technique in which numerous baited hooks are trailed from a line that can be kilometres long, has caused many seabirds, including most species of albatross, to decline rapidly in numbers.

Albatrosses follow longlining ships to feed on the bait put on the lines' hooks. The birds get hooked and then drown.

Accidentally or deliberately introduced species have been the cause of some quick and unexpected extinctions. For example, in 1960 the unique birds of the island of Guam appeared to be in no danger, for many species were quite common. Yet a reptile, the brown tree snake (*Boiga irregularis*), had been accidentally introduced perhaps a decade earlier, and, as it spread across the island, it systematically exterminated all the island's land birds. The snakes also occasionally stow away in cargo leaving Guam. And because there is substantial air traffic from Guam to Honolulu, Hawaii, some snakes arrived there. None are thought to have survived, but, should the snake establish a population there, the Hawaiian Islands would likely lose all their remaining native birds.

A wildlife specialist holding a brown tree snake (Boiga irregularis) *that was captured on a military base in Guam as part of a program to prevent the species from spreading to Hawaii. After the species was accidentally introduced to Guam in the mid-20th century, it colonized the island and preyed relentlessly on native species, causing the decline, local elimination, or extinction of several species of birds, fruit bats, and lizards.* Master Sgt. Lance Cheung/ U.S. Air Force

Although anticipating the effect of introduced species on future extinctions may be impossible, it is fairly easy to predict the magnitude of future extinctions from habitat loss, a factor that is simple to quantify and that is usually cited as being the most important cause of extinctions. (For birds, to give an example, some three-fourths of threatened species depend on forests, mostly tropical ones that are rapidly being destroyed.) Habitat destruction is continuing and perhaps accelerating, so some now-common species certainly will lose their habitat within decades. Even so, making specific predictions requires a more-detailed understanding of the factors that cause extinctions.

The Loss of Populations

While much of the concern over the loss of biodiversity centres on the global loss of species, most of the benefits that biodiversity confers depend on the continued existence of sufficient numbers of the local populations of species. An obvious example is loss of the forests that protect a city's watershed. In preventing soil erosion, these forests and their species provide a local service that would cease if the forests were cleared, irrespective of whether their species still existed elsewhere in the world. In addition, local populations supply genetic diversity, because different populations across a species' range differ to varying degrees in their genetic composition. Such genetic diversity has significant value for agricultural plant breeders, who rely on wild relatives of crop plants for novel genes that confer resistance to plant pathogens such as viruses and fungi. Thus, as populations are eliminated locally, potentially beneficial genes may become lost globally.

Scientists estimate than an average species consists of 220 genetically distinct populations, suggesting that there may be more than 2 billion such populations globally. They estimate that 160 million populations (8 percent) are lost each decade. If this trend continues, even though many species may be saved in protected areas (such as national parks), each of them will be merely a remnant of a once geographically extensive and genetically diverse set of populations.

THE LOSS OF ECOSYSTEMS

Conservation justifiably prioritizes tropical moist forests because they hold such a large fraction of the global total of species, but a comprehensive strategy must also include the conservation of other distinctive ecosystems. For example, major habitat types such as tropical dry forests, tundra, temperate grasslands, lakes, polar seas, and mangroves all contain characteristic species. Importantly, their biological communities house distinctive ecological and evolutionary phenomena. Some of these ecosystems are more threatened than are tropical moist forests and require immediate conservation action.

To provide a more representative selection of Earth's distinctive ecosystems, scientists working in the United States at the World Wildlife Fund (WWF) have compiled what they call the world's "ecoregions," which are drawn from the planet's terrestrial, freshwater, and marine major habitat types. Based on a comparative global analysis, they have identified some 240 ecoregions as global priorities—about 140 terrestrial, 50 freshwater, and 40 marine areas that are the most biologically outstanding, either for their richness of species or for their ecological or evolutionary uniqueness.

A comparison of the WWF's list of global priorities and a list of priorities based on threatened species alone uncovers many overlaps, predominantly in the tropical moist forests. It is the differences that are interesting, however. The Everglades of Florida and Brazil's Pantanal, for example, do not rank as places with high concentrations of threatened species, but they achieve prominence among the WWF's global ecoregions because they are among the few flooded grasslands in the world, all of which are under threat. Other threatened ecosystems identified on WWF's list include the dry forests of lowland Bolivia, southern Mexico, Madagascar, and Hawaii. The temperate grasslands of North America, another global priority, once sustained vast migrations of large vertebrates such as bison, pronghorn, and elk. Such species are now restricted to small pockets. The same is true in Eurasia, where comparable migrations now occur only in isolated pockets, primarily in the Daurian Steppe in China, Mongolia, and Russia and the Tibetan Plateau. And the extraordinary floral communities of the Eurasian steppes and the North American Great Plains have been largely destroyed through the land's conversion to agriculture.

CHAPTER 2

The Causes of Extinction

In theory, any factor capable of causing mortality can cause a species to die out if it is particularly intense and occurs over a wide geographic scale. Habitat loss, pollution, overharvesting, secondary extinctions, and global change may contribute to the demise of a species either singly or in concert with one another. The species most vulnerable to these phenomena include rare species, species that possess small home ranges, species whose mating systems result in inbred populations, and species that do not disperse widely.

HABITAT LOSS

As mentioned above, habitat loss is widely listed as the predominant cause of extinction. It is not hard to understand why—clear a forest, destroy a coral reef, or dam a river, and the species found there will likely be lost. These are instances of local extinctions, however, and their occurrence does not mean that the species involved will go extinct everywhere, as is the situation for the global extinctions discussed in the case histories above. Some scientists use the term *extirpation* for local extinctions, reserving *extinction* to mean global extinction. In this section on factors causing extinction, the term has the global meaning.

Extinction in Terrestrial Environments

In the preceding case histories, the roles of habitat loss in extinction of species are fairly clear-cut. There are seeming counterexamples—Europe and eastern North

America, for example—where human actions have extensively modified terrestrial habitats, yet these areas are not centres of extinction. Clearly, habitat destruction causes different numbers of extinctions in different places.

Endemic Species

To learn what makes centres of human-caused extinctions special, one can ask what are their common features. Obviously, many of these places and their species are well known. The 17th-century Dutch artist Rembrandt van Rijn painted birds-of-paradise and marine cone shells gathered from the early European exploration of the Pacific, testifying to people's fascination for attractive and interesting specimens from "exotic" locales. Victorians filled their cabinets with such curiosities—birds, mammals, marine and terrestrial snail shells, and butterflies—and painted tropical flowers and grew them in their hothouses. Yet the natural history of North America and Europe are also very well known. In fact, what is special about the places with so many extinctions is that each area holds a high proportion of species restricted to it. Scientists call such species endemics.

Remote islands have many terrestrial endemics—for instance, more than 90 percent of the plants and land birds of the Hawaiian Islands live nowhere else. Some continental areas are rich in endemic species, too. About 70 percent of the flowering plants in South Africa's fynbos and nearly three quarters of Australia's mammals are endemic. In contrast, many areas have almost no endemic species. Only about 1 percent of Europe's birds are found only there.

The simplest model of extinction would be to assume that within a species group—mammals, for instance—all species had roughly the same risk of extinction. Were this

Many Victorians were fascinated by paintings of "exotic" subjects and places. Woman with Parrot, *painting on copper by the Dutch artist Frans van Mieris the Elder (1635–81); in the National Gallery, London.* Courtesy of the trustees of the National Gallery, London; photograph, A.C. Cooper

Hawaiian honeycreepers such as the Apapane (Himatione sanguinea) *are considered endemic.* Drawing by John P. O'Neill

to be true, then the more species that live in a region, the more would likely go extinct. This is not the case, however. For example, the Hawaiian Islands and Great Britain in their pasts held very roughly the same number of breeding land birds. Yet, the former have lost more than 100 species, as discussed in the preceding Pacific Islands case study, while the latter has lost only a few—and those still survive on the European continent. The difference is that almost all the Hawaiian birds (for instance, honeycreepers such as the apapane and iiwi) were endemic, while only one of Great Britain's birds (the Scottish crossbill) is. Thus, the number of extinctions in an area depends only very weakly on its total number of species but strongly on its total number of endemics. Areas rich in endemic species are where extinctions will concentrate, unless they are so remote that human actions do not harm them.

HAWAIIAN HONEYCREEPER

Any member of a group of related birds, many of them nectar-eating, that evolved in the forests of the Hawaiian Islands and are found only there are called Hawaiian honeycreepers. Recent evidence from osteology, behaviour, plumage, breeding biology, and genetics has led to a consensus that the Hawaiian honeycreepers are closely related to the cardueline finches, which include birds such as goldfinches, canaries, siskins, and crossbills. They constitute the family Drepanididae within the order Passeriformes. Most of the species are called by native names (such as amakihi, apapane, iiwi, and mamo). Habitat destruction and the introduction of foreign birds and mammals have led to the extinction of at least 8 of the original 23 species; most of the survivors are endangered. Numerous subspecies are known.

Hawaiian honeycreepers differ in certain ways from American honeycreepers. Isolated in the mid-Pacific, they underwent a remarkable evolutionary radiation, diversifying in the manner of the better-known Darwin's finches in the Galapagos Islands. Those with thin bills and, usually, red-and-black plumage (both sexes look alike) feed on nectar; those with finchlike bills and, usually, greenish plumage (males often have orange or yellow markings) eat seeds, fruits, and insects. Other species are intermediate between these two types. In most Hawaiian honeycreepers the tongue is troughlike and brush-tipped. The birds' size ranges from 10 to 20 cm (4 to 8 inches). Hawaiian honeycreepers usually have simple songs and make grassy nests.

The Iiwi (Vestiaria coccinea) *is one of many Hawaiian honeycreepers that have radically diversified.* M. Ord/Photo Researchers

Range Size

Given that species differ in their risk of extinction, the size of a species' geographic range is by far the best explanation for the differences. Species with small ranges are much more vulnerable than those with large ranges, simply because it is much easier to destroy the former than the latter.

In regard to range size, the distribution of life on land has several remarkable features. First, many terrestrial species have very small range sizes relative to the average range size. In the Americas, for example, 1 in 10 species of birds and over half the species of amphibians have geographic ranges smaller than the state of Connecticut, and half the bird species have ranges smaller than the states of Washington, Oregon, and California combined. The average range size is much larger, for some species have huge ranges. The American robin (*Turdus migratorius*), for example, breeds almost everywhere in the United States from Alaska to Florida to California, across all of continental Canada, and in much of Mexico.

Second, for many kinds of terrestrial organisms, species with small ranges typically have lower local population densities than do widespread species. For example, the American robin is generally a locally common bird across its entire range. But those species with ranges smaller than the size of Connecticut are generally difficult to find even in the midst of their ranges.

Third, terrestrial species with small ranges are geographically concentrated. North America, for instance, has few bird species with small ranges. Such species live almost exclusively in the tropics. The greatest concentrations of bird species are in the Amazon lowlands, with secondary centres in Central America and the forests along the coast of Brazil. But it is the Andes and the coastal forests that have the most bird species with small geographic ranges (i.e., these areas are the centres of endemism). And it is in these areas that threatened species are concentrated.

Terrestrial Hot Spots

In the 1990s a team of researchers led by British environmental scientist Norman Myers identified 25 terrestrial "hot spots" of the world: 25 areas on land where species with small geographic ranges coincide with high levels of modern human activity. Originally, these hot spots encompassed about 17 million square km (6.6 million square miles) of the roughly 130 million square km (50 million square miles) constituting Earth's ice-free land surface. Species ranges are so concentrated geographically in these regions that, out of a total of about 300,000 flowering-plant species described worldwide, more than 133,000 occur only there. The comparable numbers for birds are roughly 2,800 of 10,000 species worldwide (of which roughly two-thirds are restricted to the land); for mammals, 1,300 of roughly 5,000 worldwide; for reptiles, roughly 3,000 of about 8,000 worldwide; and for amphibians, 2,600 of roughly 5,000 worldwide.

The hot spots have been sites of unusual levels of habitat destruction. Only about one-eighth of the original habitat of these areas has survived to the beginning of the 21st century. Of this remaining habitat, only about two-fifths is protected in any way. Sixteen of the 25 hot spots are forests, most of them tropical forests. For comparison, the relatively less disturbed forests found in the Amazon, the Congo region, and New Guinea have retained about half their original habitats. As a consequence of these high levels of habitat loss, the 25 hot spots are locations where the majority of threatened and recently extinct species are to be found.

Fire Suppression as Habitat Loss

Whereas most hot spots are tropical moist forests, four areas are shrublands: the California Floristic Province, the Cape Floristic Province in South Africa, the Mediterranean Basin, and Southwest Australia. They also are places where people live and grow crops. All four regions are noted for their wines, for example. Not only does this human activity convert land directly, but it also leads to the suppression of fire, especially near people's homes. This alteration of natural fire regimes by the reduction in fire frequencies leads to changes in vegetation, especially to the loss of the native fire-resistant species. Globally, huge areas of grasslands and shrublands would become heavily canopied forests were all fires suppressed. The effects of changes in fire frequencies on species losses have not yet been calculated.

The Predictions of Extinctions Based on Habitat Loss

Worldwide, about 6 percent of the land surface is protected by some form of legislation, though the figure for the 25 hot spots is only 4.5 percent of their original extent. (Such numbers are misleading, however, in that some areas are protected only on paper as their habitats continue to be destroyed.) These statistics lead naturally to the question of how many species will be saved if, say, 4.5 percent of the land worldwide is protected. Will only 4.5 percent

The coastal forest of Rio de Janeiro state, Braz., badly fragmented as portions were cleared for cattle grazing. Courtesy, Stuart L. Pimm

of its species be saved? The answer turns out to be closer to 50 percent, a result that needs some explanation.

The land that countries protect, often as national or regional parks, frequently comprises "islands" of original habitat surrounded by a "sea" of cropland, grazing land, or cities and roads. On a real island the number of species that live there depends on its area, with a larger island housing more species than a smaller one. Many studies involving a wide range of animals and plants show that the relationship between area and species number is remarkably consistent. An island half the size of another will hold about 85 percent of the number of species.

The next question is whether this relationship of species to area holds for "islands" of human-created habitat. If one were to conduct an experiment to test such an idea, one would take a continuous forest, cut it up into isolated patches, and then wait for species to become locally extinct in them. After sufficient time, one could count the numbers of species remaining and relate them to the area of the patches in which they survived. In the last decades of the 20th century, scientists undertook exactly such an experiment, making use of government-approved forest clearing for cattle ranching, in the tropical forests around Manaus in Brazil. More generally, human actions have repeated this experiment across much of the planet in an informal way. Counts of species in areas of different sizes confirm the species-to-area relationship.

How can these results regarding local habitat "islands" be applied to global extinctions? The mostly deciduous forests of eastern North America provide a case history. The birds of the region have been well-described beginning with the explorations of the naturalist and artist John James Audubon in the first quarter of the 19th century, when the area was still mostly forested. Audubon shot and painted many species including four that are now extinct: the Carolina parakeet (*Conuropsis carolinensis*), Bachman's warbler (*Vermivora bachmanii*), the passenger pigeon (*Ectopistes migratorius*), and (if not extinct, then very nearly so) the ivory-billed woodpecker (*Campephilus principalis*). A fifth species, the red-cockaded woodpecker (*Picoides borealis*), is threatened. Including those just listed, about 160 species of birds once lived in these eastern forests.

European settlement cleared these forests, and, at the low point about 1870, only approximately half the forest

The passenger pigeon (Ectopistes migratorius) *is just one of about 160 species of North American birds that once flourished in the eastern forests.* Encyclopædia Britannica, Inc.

remained. The species-to-area relationship predicts that about 15 percent of the 160 bird species—that is, about 24 species—would become extinct. Why is this number much larger than the three to five extinctions or near-extinctions observed? The answer can be found by first supposing that all the eastern forests had been cleared, from Maine to Florida and westward to the prairies. Only those species that lived exclusively within these forests—that is, the endemics—would have gone extinct. Species having much larger ranges, such as the American robin mentioned earlier, would have survived elsewhere. How many species then were originally endemic to the forests of eastern North America? The answer is about 30, the rest having wide distributions across Canada and, for some, into Mexico. The species-to-area predicts that 15 percent of these 30 species, or 4.5 species, should go extinct, which is remarkably close to the observed number.

Eastern North America clearly is not a place where species with small ranges are concentrated, but the species-to-area predictions work in other places, too. A variety of studies have examined birds and mammals on the islands of Southeast Asia, from Sumatra westward—one of the biodiversity hot spots (Sundaland). Both here and for birds in Brazil's Atlantic Forest, which is another hot spot, the extent of deforestation and the species-to-area relationship accurately predict the number of threatened species rather than extinct ones. Because the deforestation of these areas is relatively recent, many of their species have not yet become extinct.

The close match between the numbers of extinctions predicted by the species-to-area relationship and the numbers of species already extinct (as in eastern North America) or nearly extinct (as in more recently destroyed habitats) allows simple calculations that have worldwide import. As previously discussed, the hot spots retain only

12 percent of their original habitats, and only about 4.5 percent of the original habitats are protected. As predicted by the species-to-area relationship, natural habitats that have shrunk to 4.5 percent of their original extent will lose more than half of their species. Because these habitats once supported 30–50 percent of terrestrial species, very roughly one-fourth of all terrestrial species will likely become extinct.

Species losses will likely be even greater because this calculation does not include nonendemic species—those that live both inside and outside the hot spots. For example, many of the species that live in the relatively less-disturbed tropical moist forests of the Amazon or the Congo region are the same that live in the adjacent hot spots. Human actions are clearing about 10 percent of the original area of these forests every decade, with a half of the area already gone. Species losses in these forests are still relatively few, but the rate will increase rapidly as the last remaining forests dwindle. If only the same percentage of these forests is protected as is the case for the hot spots, then they too will lose half their species.

In summary, many scientists believe that habitat destruction will put somewhere between a fourth and a half of all species on an inexorable path to extinction and will do so within the next few decades. If that proves true, extinction rates by the mid-21st century will be several thousand times the benchmark rate.

Extinction in Marine Environments

The seas cover more than two-thirds of Earth's surface, yet only 210,000 of the 1,500,000 species that have been described are marine animal and plant species. Because the oceans are still poorly explored, the count of marine species may be even more of an underestimate than that of land species. For example, the Census of Marine Life, a decade-long international program begun at the start of the 21st century, added 13,000 new marine species to the total count over the first four years of the effort. As on land, the peak of marine biodiversity lies in the tropics. Coral reefs account for almost 100,000 species, yet their total area represents just 0.2 percent of the ocean surface. Between 4,000 and 5,000 species of fish—perhaps a third of the world's marine fish—live on coral reefs. The frequently cited metaphor that "coral reefs are the rainforests of the sea" underscores their importance for conservation.

When numbers of described marine species are mapped on a worldwide scale, it becomes clear that the global centre of marine biodiversity encompasses the waters of the Philippine and Indonesian islands. Numbers of species drop steeply to the east across the Pacific and less steeply to the west across the Indian Ocean. In the Atlantic Ocean the highest levels of biodiversity are in the Caribbean. Fish, corals, mollusks, and lobsters all show similar patterns in the distributions of their species. Again mirroring the patterns on land, the places with the most species are often not the places with the most endemic species. Major centres of endemism for fish, corals, mollusks, and lobsters include the Philippines, southern Japan, the Gulf of Guinea, the Sunda Islands, and the Mascarene Islands.

With the major exception of the Great Barrier Reef of Australia, most coral reefs are off the coasts of developing countries. Rapidly increasing human populations and poverty put increasing fishing pressure on nearshore reefs. In addition, in their efforts to sustain declining fish catches, people resort to extremely damaging fishing methods such as dynamite and poisons. Coral reefs are also threatened by coastal development, pollution, and global warming. Human activities threaten some three-fifths of the world's reefs, with the highest damage being concentrated in areas having high rates of deforestation and high runoff from the land. As the inhabitants of an area destroy their tropical forests, rains erode soils and wash the sediments down rivers into the sea, damaging the local coral reefs. Thus, the destruction of some of the most important terrestrial habitats—in the Caribbean and Southeast Asia—contributes to the destruction of some of the most important marine habitats offshore.

Whereas damage to coral reefs is important for the loss of species, the greatest physical damage to ocean ecosystems involves the effects of bottom trawling, a commercial fishing method making use of a cone-shaped bag of netting that is dragged along the seabed. Damage from bottom trawling occurs over larger areas of Earth than does tropical deforestation, and it involves even greater and more-frequent disturbances, albeit ones not easily seen. Bottom trawling disturbs about 15 million square km (6 million square miles) of the world's seafloor each year. This area of ocean is only about 4 percent of the world total, but its small proportion belies its significance. About 90 percent of the ocean consists of deep waters so poor in nutrients that they are the equivalent of the land's deserts. Almost all of the world's fisheries are concentrated in the 30 million square km (12 million square miles) of nutrient-rich waters that are on the continental shelf, plus a few upwellings. On average, the ocean floor of these productive waters is trawled roughly every two years.

The otter trawl is the most widely used bottom-fishing gear. As it is dragged forward, a pair of flat plates called otter boards—one on each side of the trawl net and weighing several tons—spreads horizontally to keep the mouth of the trawl open; at the same time, a long rope with steel weights keeps the mouth open along its bottom edge. This heavy structure plows the ocean floor as it moves, creating furrows and crushing, burying, and exposing marine life. This activity destroys bottom-dwelling species including corals, brachiopods, mollusks, sponges, sea urchins, and various worms that live on rocks or pebbles on the seabed. It is also damaging to other species, such as polychaete worms that burrow into the seabed. Some species—deep-sea corals, for example—are extremely slow-growing, and they cannot recover before bottom trawls plow the area once again.

Extinction in Freshwater Environments

Freshwater ecosystems are divided into two major classes—flowing (such as rivers and streams) and static (such as lakes and ponds). Although the distribution of species in freshwater ecosystems is not as well-known as for marine and terrestrial ecosystems, it is still clear that species are similarly concentrated. For fish, the major tropical rivers such as the Amazon River and its tributaries hold a large fraction of the world's freshwater fish species. Tropical lakes, particularly those in the Rift Valley of East Africa, also have large numbers of endemic species.

Riverine habitats have been extensively modified by damming and by channelization, the latter being the practice of straightening rivers by forcing them to flow along predetermined channels. A global survey published in the early 21st century revealed just how few of the world's large rivers are natural. In the contiguous United States, almost every large river has been modified extensively. (As previously noted, some rivers such as the Tennessee River have been converted almost completely by dams to a series of artificial lakes.) Much the same is true in Europe. In terms of the area of their basins, more than half of the world's rivers are extensively modified. The water of some rivers barely reaches its final destination. Such is the case for the Colorado River in the United States, which empties into the Gulf of California, and for the Amu Darya in Central Asia, which empties into the rapidly shrinking Aral Sea. Along their routes the water is used for agriculture or lost as evaporation from dams. Large wild rivers are typical only of Arctic regions in Alaska, Canada, and Siberia—places so far away from urban centres that there has been no incentive to control their waters. The massive changes to the world's rivers explain why such large fractions of species living in rivers have become extinct or may do so soon (as is described in two of the preceding case histories).

POLLUTION

Pollution is a special case of habitat loss that involves chemical destruction rather than the more obvious physical destruction. Pollution occurs in all habitats—land, sea, and fresh water—and in the atmosphere. Global warming is one consequence of the increasing pollution of the atmosphere by emissions of carbon dioxide and other gases.

Pollution is a global-scale problem, no less so for rivers and marine life. Wastes are often dumped into rivers, and they end up in estuaries and coastal habitats, regions that support the most diverse shallow-water ecosystems and the most productive fisheries. Rivers receive pollution directly from factories that dump a wide variety of wastes into them. They also receive runoff, which is rainwater that has passed over and through the soil while moving toward the rivers. In fact, water entering rivers after it has been used for irrigation has passed through the soil more than once—first as runoff, which is then returned to the land for irrigation, whereupon it soaks through the soil again.

Some polluted river water eventually reaches freshwater wetlands. In the case of the Florida Everglades, runoff from the agricultural areas upstream adds unwanted nutrients to an ecosystem that is naturally nutrient-poor. As it does so, the vegetation changes, and species not common in the Everglades begin to take over the natural habitats.

Other polluted waters reach estuaries on the way to the oceans. Estuaries are among the most polluted ecosystems on Earth. On entering the oceans, the polluted waters can harm the ecosystems there. The Mississippi River, for example, drains a basin of more than 3 million square km (1.2 million square miles), delivering its water, sediments, and nutrients and other pollutants into the northern Gulf of Mexico. Fresh water is less dense than salt water and floats on top. This upper layer contains the nitrogen and phosphorus fertilizers that have run off croplands, and they fertilize the ocean's phytoplankton, causing excessive population growth. As the masses of phytoplankton die, sink, and decompose, they deplete the water's oxygen. Bottom dwellers such as shrimp, crabs, starfish, and marine worms suffocate and die, creating a "dead zone."

Such conditions also occur in Europe's Baltic, Adriatic, and Black seas. The Baltic has gone from being naturally nutrient-poor and diverse in species to being nutrient-rich and degraded in its ecosystems within a few decades. In the Adriatic Sea, rising nutrient levels have generated a large increase in phytoplankton. Nutrients in the runoff flowing into the Black Sea seem to be contributing factors in the invasion and subsequent massive increase since the 1980s of the comb jelly *Mnemiopsis leidyi*. This has caused the decline of native species and fisheries.

Similar nutrient enrichment has led to increasing frequencies of toxic blooms of microscopic organisms such as *Pfiesteria piscicida* in the eastern United States, a dinoflagellate that kills fish and has been reported to cause skin rashes and other maladies in humans.

Rising levels of pollution may have also contributed to a wave of outbreaks of diseases affecting marine life. Caribbean coral reefs have been particularly affected, with successive waves of disease propagating throughout the region in recent decades. The result has been large declines in two species of major reef-building corals, *Acropora cervicornis* and *A. palmata*, and the herbivorous sea urchin *Diadema antillarum*. Their combined loss has transformed Caribbean reefs from high-coral, low-algae ecosystems to high-algae, low-coral ones. The latter type of ecosystems support far fewer species.

THE ROLE OF INTRODUCED SPECIES

The case histories previously discussed often implicate introduced species as a cause of species extinctions. Humans have spread species deliberately as they colonized new areas, just one example being the Polynesians as they settled the eastern Pacific Islands. New Yorkers in the 1890s wanted all the birds in Shakespeare's works to inhabit the city's Central Park, and they introduced the starling (*Sturnus vulgaris*) to North America as a consequence. Through the centuries hunters have demanded exotic birds and mammals to shoot, fishermen have wanted challenging fish, and gardeners have wanted beautiful flowers. Nonetheless, the consequences in some cases have been devastating. Cacti and the shrub *Lantana camara*, for example, which were introduced as ornamental plants, have destroyed huge areas of grazing land worldwide.

The common starling (Sturnus vulgaris) *was introduced to New York City by 19th-century Shakespeare enthusiasts who wanted all the bard's birds represented in Central Park.* George W. Robinson/Root Resources

Not all introductions are deliberate. Several species of rats, for instance, were likely unwelcome hitchhikers on ocean voyages. Humans have spread infectious diseases around the world with consequences that dramatically changed human history. A recent example is AIDS, spread from Africa by individuals infected with the human immunodeficiency virus (HIV). Introduction of the tree snake that destroyed Guam's birds was also accidental.

Almost every type of living organism has been deliberately or accidentally moved. A study published in the 1990s surveyed species that traveled in the ballast tanks of oceangoing ships. Such ships take on seawater as ballast in one port and release it in other ports sometimes half a world away. In one example, water taken aboard in Japan and released in Oregon contained 360 species including microscopic single-celled animals and plants, jellyfish and corals, mollusks, various kinds of marine worms, and fish—indeed, an almost complete catalog of major kinds of marine organisms.

Although not all species devastate the communities they enter, a closer look at what happened on Guam after the introduction of the brown tree snake (*Boiga irregularis*) illustrates just how damaging a species can be. Birds started disappearing from the central region of the island in the early 1960s. At the same time, the snakes started causing blackouts as they crawled up power poles and short-circuited the suspended lines, and many people who kept chickens noticed that snakes were killing the birds in their coops. Before *Boiga* made its appearance, 10 species of native birds lived in the forests of Guam: the Micronesian starling (*Aplonis opaca*), the Mariana crow (*Corvus kubaryi*), the Micronesian kingfisher (*Halcyon cinnamomina*), the Guam flycatcher (*Myiagra freycineti*), the Guam rail (*Rallus owstoni*), the rufous fantail (*Rhipidura rufifrons*), the bridled white-eye (*Zosterops conspicillatus*), the white-throated ground dove (*Gallicolumba xanthonura*), the Mariana fruit dove (*Ptilinopus roseicapilla*), and the cardinal honeyeater (*Myzomela cardinalis*). An 11th species, the island swiftlet (*Aerodramus vanikorensis*), nested in caves and fed over the forest. In addition, the island supported a few introduced bird species, birds that nested in wetland habitats, and some seabirds. The 10 forest birds declined following similar patterns—first on the central part of the island, where the main town and port are, and then on southern Guam by the late 1960s. By 1983 scientists could find the birds only in a small patch of forest in the north of the island, and by 1986 they were gone from there as well. The nonnative birds suffered a similar pattern of decline, and many of the waterbirds and seabirds that nested on the island also declined. The starling has survived on a small island, Cocos, off the south coast of Guam, and the swiftlet still nests on the high walls of a cave. The snake has eliminated every species for which it could easily kill the eggs and young in the nest. Of the 10 forest species, the rail and the flycatcher were endemic to the island, while the kingfisher is a distinctive subspecies. The kingfisher and the rail survive in captivity.

Islands such as Guam seem to be particularly vulnerable to introduced species. People have often brought

The now-extinct Rufous fantail (Rhipidura rufifrons) *is just one of 10 species of birds native to Guam before the fatal introduction of the brown tree snake. E.* McNamara—Ardea, London

domestic cats with them to islands, where some of the animals have escaped to form feral populations. Cats are efficient predators on vertebrates that have had no prior experience of them. For example, cats have caused the extinction of some 30 island bird species worldwide. Introduced herbivores, particularly goats, devastate native plant communities, again because the plants have had no chance to adapt to these new threats.

Although islands are particularly vulnerable, introduced species also wreak large-scale changes on continents. An exceptionally clear example is the loss of the American chestnut (*Castanea dentata*) in eastern North America after the accidental introduction of the fungus, *Endothia parasitica*, that causes chestnut blight. The chestnut, a once-abundant tree, was removed with surgical precision over its entire range beginning in the early 20th century. All that survive are small individuals that tend to become infected as they get older.

As briefly mentioned earlier, hybridization is another mechanism by which introduced species can cause extinction. In general, species are considered to be genetically isolated from one another—they cannot interbreed to produce fertile young. In practice, however, the introduction of a species into an area outside its range sometimes leads to interbreeding of species that would not normally meet. If the resident species is extremely rare, it may be genetically swamped by the more abundant alien. One example of a species threatened by hybridization is the white-headed duck (*Oxyura leucocephala*). The European population of this species lives only in Spain, where habitat destruction and hunting once reduced it to just 22 birds. With protection, it recovered to about 800 individuals, but it is now threatened by a related species, the ruddy duck (*O. jamaicensis*). This bird is native to North

America, was introduced to Great Britain in 1949, and spread to the continent including Spain, where it hybridizes with the much rarer white-headed duck. Similar cases involve other species of ducks, frogs, fish, cats, wolves, and other groups of organisms worldwide.

OVERHARVESTING

Overharvesting, or overfishing in the case of fish and marine invertebrates, depletes some species to very low numbers and drives others to extinction. In practical terms, it reduces valuable living resources to such low levels that their exploitation is no longer sustainable. Whereas the most-familiar cases involve whales and fisheries, species of trees and other plants, especially those valued for their wood or for medicines, also can be exterminated in this way.

CITES

To prevent the overexploitation of species as they are traded across national boundaries, the Convention on International Trade in Endangered Species of Wild Flora and Fauna (CITES) was created by international agreement in 1973 and put into effect in 1975. The agreement sorts more than 5,000 animal and 28,000 plant species into three categories (denoted by its three appendixes). Appendix I of CITES lists the species in danger of extinction. Although it also prohibits outright the commercial trade of these species, some can be traded in extraordinary situations for scientific or educational reasons. In contrast, Appendix II lists particular plants and animals that are less threatened but still require stringent controls. Appendix III lists species that are protected in at least one country that has petitioned other countries for help in controlling international trade in that species. As of 2009, CITES had been signed by 175 countries.

Whaling

Whaling offers an example of overharvesting that is interesting not only in itself but also for demonstrating how poorly biodiversity has been protected even when it is of economic value. The first whalers likely took their prey close to shore. Right whales were the "right" whales to take because they are large and slow-moving, feed near the surface and often inshore, float to the surface when harpooned, and were of considerable commercial value for their oil and baleen. The southern right whale (*Eubalaena australis*), for example, is often seen in shallow, sheltered bays in South Africa and elsewhere. Such behaviour would make any large supply of raw materials a most tempting target. Whalers had nearly exterminated the North Atlantic species of the northern right whale (*Eubalaena glacialis*) and the bowhead whale (Greenland right whale; *Balaena mysticetus*) by 1800. They succeeded in exterminating the Atlantic population of the gray whale (*Eschrichtius robustus*). Whalers then moved on to species that were more difficult to kill, such as the humpback whale (*Megaptera novaeangliae*) and the sperm whale (*Physeter macrocephalus*).

The Napoleonic Wars gave whales a respite, but with the peace of 1815 came a surge of whalers into the Pacific Ocean, inspired by the stories of James Cook and other explorers. The first whalers arrived in the Hawaiian Islands in 1820, and by 1846 the fleet had grown to nearly 600 ships, the majority from New England. The catch on each whaling voyage averaged 100 whales, though a voyage could last as long as four years.

In the late 1800s, steamships replaced sailing ships, and gun-launched exploding harpoons replaced hand-thrown lances. The new technology allowed whalers to kill what until then had been the "wrong" whales—fast-swimming species such as the blue whale (*Balaenoptera musculus*) and fin whale (*B. physalus*). Whalers killed nearly 30,000 blue whales in 1931 alone; World War II gave the whales a break, but the catch of blue whales rose to 10,000 in 1947. The fin whale was next, with the annual catch peaking at 25,000 in the early 1960s; then came the smaller sei whale (*B. borealis*)—which no one had bothered to kill until the late 1950s—and finally the even smaller minke whale (*B. acutorostrata*), which whalers still hunt despite an international moratorium in effect since 1986 that seeks to curb commercial whaling.

The story of whaling is, in brief, the rapid depletion and sometimes extermination of one population after another, starting with the easiest species to kill and progressing to the most difficult. That whales are economically valuable raises the obvious question of why there were no attempts to harvest whales sustainably.

Fishing

Overfishing is the greatest threat to the biodiversity of the world's oceans, and contemporary information published for fisheries in the United States can serve as an example of the magnitude of the problem. Congress requires the National Marine Fisheries Service (NMFS) to report regularly on the status of all fisheries whose major stocks are within the country's exclusive economic zone, or EEZ. (Beyond its territorial waters, every coastal country may establish an EEZ extending 370 km [200 nautical miles] from shore. Within the EEZ the coastal state has the right to exploit and regulate fisheries and carry out various other activities to its benefit.) The areas involved are considerable, covering portions of the Atlantic, the Caribbean, the Gulf of Mexico, and the Pacific from off San Diego to the Bering Sea out to the west of the Hawaiian island chain along with the islands constituting the western part of the former Trust Territory of the Pacific Islands. At the turn of the 21st century, the NMFS deemed some 100 fish stocks to be overfished and a few others close to being so, while some 130 stocks were not thought to be overfished. For about another 670 fish stocks, the data were insufficient to allow conclusions. Thus, a little under half of the stocks that could be assessed were considered overfished. For the major fisheries—those in the Atlantic, the Pacific, and the Gulf of Mexico—two-thirds of the stocks were overfished.

As to the hundreds of stocks about which fisheries biologists know too little, most of them are not considered economically important enough to warrant more investigation. One species, the barn-door skate (*Raja laevis*), was an incidental catch of western North Atlantic fisheries in the second half of the 20th century. As the name suggests, this is a large fish, too big to go unrecorded. Its numbers fell every year until, by the 1990s, none were being caught.

Logging and Collecting

Similar cases of overharvested species are found in terrestrial ecosystems. For example, even when forests are not completely cleared, particularly valuable trees such as mahogany may be selectively logged from an area, eliminating both the tree species and all the animals that depend on it. Another example is the coast sandalwood (*Santalum ellipticum*), a tree endemic to the Hawaiian Islands that was almost completely eliminated from its habitats for its wood and fragrant oil.

Some species are overharvested not to be killed but to be kept alive and sold as pets or ornamental plants. Many species of parrots worldwide, for example, are in danger because of the pet trade, and the survival of cacti and orchid species is threatened by collectors.

SECONDARY EXTINCTIONS

Once one species goes extinct, there will likely be other extinctions or even an avalanche of them. Some cases of these secondary extinctions are simple to understand (e.g., for every bird or mammal that goes extinct, one or several species of parasite also will likely disappear). From well-studied species, it is known that bird and mammal species tend to have parasites on or inside them that can live on no other host.

Other extinctions cause changes that can be quite complicated. Species are bound together in ecological communities to form a food web of species interactions. Once a species is lost, those species that fed on it, were

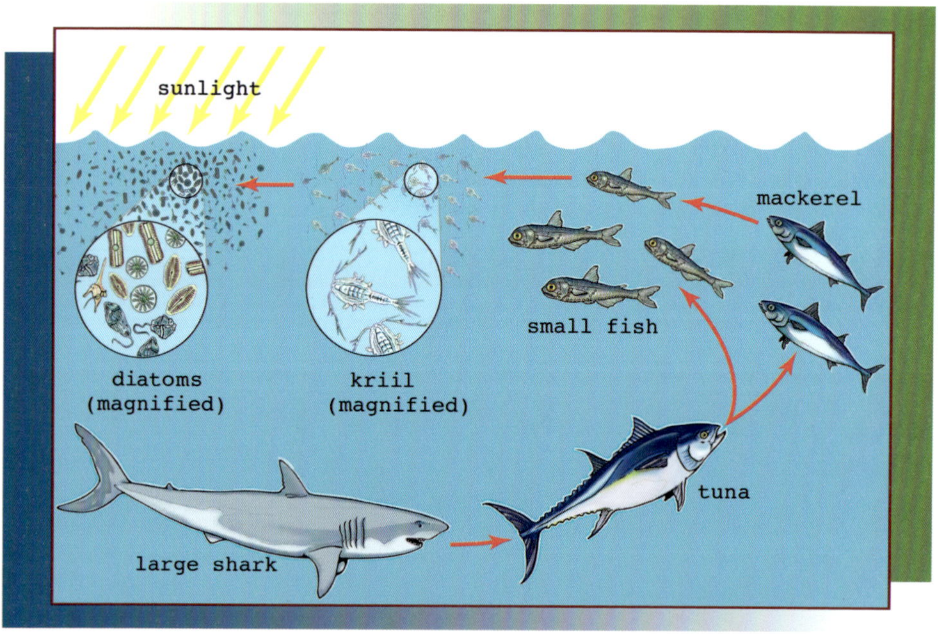

Diatoms and other phytoplankton form the foundations of ocean food chains. Shrimplike krill consume the phytoplankton, and small fishes eat the krill, and so on. One animal's extinction can upset the entire balance of the food chain. Encyclopædia Britannica, Inc.

fed on by it, or were otherwise benefited or harmed by that species will all be affected, for better or worse. These species, in turn, will affect yet other species. Ecological theory suggests that the patterns of secondary extinction are quite complicated and thus may be difficult to demonstrate.

The most easily recognizable secondary extinctions should be seen in species that depend closely on each other. In the Hawaiian Islands, anecdotal evidence for secondary extinctions comes from a consideration of nectar-feeding birds. Before modern human activity on the islands, there were three nectar-feeding Hawaiian honeycreepers—the mamo (*Drepanis pacifica*), the black mamo (*Drepanis funerea*), and the iiwi (*Vestiaria coccinea*)—that had long decurved (downward-curving) beaks, the kind adapted to inserting into appropriately long and curved flowers. The first two birds are extinct, whereas the third is extinct on two islands, is extremely rare on a third, and has declined on others.

Extensive habitat destruction is likely the cause of the extinction of many species of Hawaiian birds, as previously discussed. In addition, native Hawaiians hunted some species for their feathers. In the case of the three previously described honeycreepers, however, their extinctions may have followed the destruction of important nectar-producing plants by introduced goats and pigs. Many of the native lobelias, such as those of the genera *Trematolobelia* and *Clermontia*, have clearly evolved to be pollinated by the three honeycreepers, and the plants were once important components of the forest's understory. About a quarter of these plant species are now extinct, a rate that plainly exceeds those for the rest of the flora, perhaps because they were so vulnerable to introduced mammalian herbivores. It is not certain, however, whether the plants disappeared first and then their bird pollinators or vice versa.

The Mamo (Drepanis pacifica)*, a nectar-feeding honeycreeper, thrived on the Hawaiian Islands, perhaps until the destruction of important nectar-producing plants by introduced livestock.* Painting by H. Jon Janosik

Similarly, some surviving Hawaiian birds seem to be unusually specialized feeders and to be threatened as a consequence of the loss of their food sources. For example, another rare honeycreeper, the akiapolaau (*Hemignathus munroi*), is an insectivore that feeds on insects mainly on large koa (*Acacia koa*) trees. Today, however, few koa forests remain, because the trees have been overharvested for their attractive wood. Yet another Hawaiian honeycreeper, a seed-eating species called the palila (*Loxioides bailleui*), is endangered because it depends almost exclusively on the seeds of one tree, the mamane (*Sophora chrysophylla*), which is grazed by introduced goats and sheep.

Stories of secondary extinctions are nearly always unsatisfactory anecdotes because of the difficulty of teasing apart the various explanations for what happened in the past. There is abundant evidence from small-scale ecological experiments that a change in one species' numbers (including its complete elimination) will cause cascading effects in the abundance of other species. The stories are plausible enough that particular attention should be paid to the future consequences of contemporary extinctions.

GLOBAL CHANGE

The global effects of flooding the atmosphere with carbon dioxide and other greenhouse gases created as by-products of human activity are many and complex. Global warming, an increase in global average surface temperature, is but one of them. Both the land and the oceans are increasing in temperature at different rates in different places. The Arctic and Antarctic seem to be heating up the most, while temperature changes in the tropics are more modest. In addition, the heating is melting glaciers and ice sheets, causing sea levels to rise, and it may be increasing the frequency of the most intense hurricanes and perhaps the intensity of other storms and extreme weather, with consequences to the flow of rivers. Some areas and their

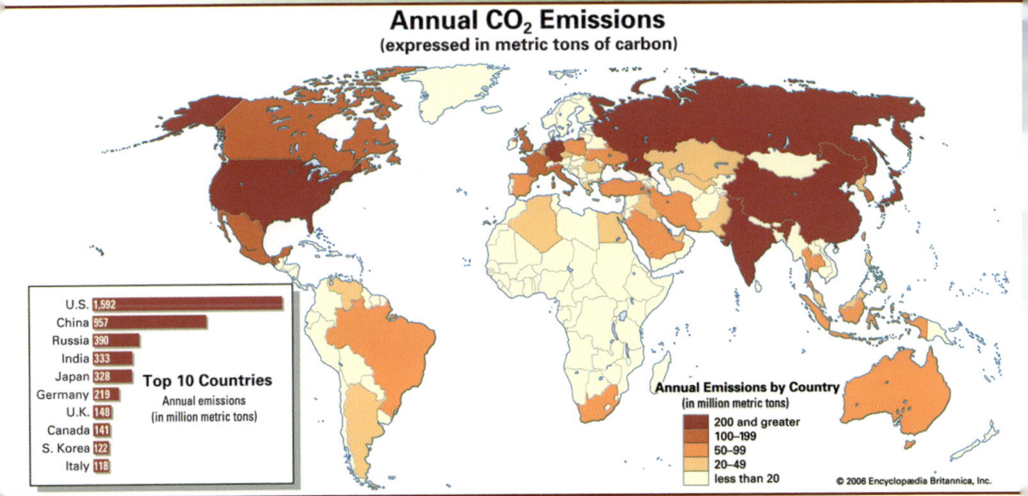

ecosystems are becoming wetter, while others are becoming drier and thus likely to suffer more wildfires. Still other areas may even become colder. Scientific knowledge is constantly increasing and changing what is known about the way Earth works, and knowledge of global change and its effects on species is expanding particularly quickly.

Since the turn of the 21st century, the relationship between changes to Earth's physics and chemistry and biodiversity has been clarified significantly. Although the precise effects of global change on species extinction rates are still uncertain, they almost certainly will be large.

It is now clear that most species are shifting their geographic ranges toward cooler places and are starting important events such as breeding, migration, and flowering earlier in the year. One study, published in the early

Some areas are becoming drier, which increases the chances of fires such as 2007's Zaca wildfire in Santa Barbara county in southwestern California. John Newman/U.S. Forest Service

21st century, found that more than 80 percent of nearly 1,500 species of animals and plants from a wide variety of habitats worldwide were changing in the direction expected from global warming. Another study, published about the same time, found consistent northward movement of the northern boundaries of animal and plant species in Europe and North America. Such geographic shifts of species alter important ecological interactions with their prey, predators, competitors, and diseases. Some species can benefit. Others can be harmed, such as when migratory insect-eating birds arrive too late to exploit the emergence of moth larvae that they typically feed to their young.

Even if scientists know that species likely will move in the direction of cooler habitats—toward the poles or up mountainsides—they will find the exact changes difficult to predict because so many different factors are involved. For example, one contemporary study examined the changes in the ranges and feeding ecology of butterflies in Great Britain in the last decades of the 20th century. Two species were found to have increased in range, as expected, but by unexpected amounts. One species lived in isolated habitat patches and could not fly the distance to some of the patches that were suitable for populating. Rising temperatures increased the number and density of suitable patches, allowing the butterfly to reach the distant patches by making use of intervening newly available ones as "stepping-stones." The second butterfly species was able to exploit a previously unused food plant that grew in shady places that formerly had been too cool. With that change in diet, the butterfly was able to greatly expand its range. In both cases, the species benefited from the warming climate, at least at the northern edges of their ranges. It takes little imagination to foresee the consequences if, for example, the mosquito *Aedes aegypti*, which carries dengue hemorrhagic fever, were to expand its range to an unexpected degree across the southern United States, which is its present northern limit.

Can anything useful be inferred about the likely consequences of climate change to species extinction? Assuming that the change has simple effects, scientists can predict where a species should be in the future if it is to live in the same range of climatic conditions that it does now. The real concern is that this range of suitable conditions, or the species' climate envelope, may shrink to nothing as conditions change (i.e., there may be no suitable conditions for a species in the future).

Other things being equal, one would expect that species with small geographic ranges will more likely be affected than species with large ranges. For a species with a large range, global change may cause the species to disappear from the south and appear farther north, but the change in the size of the range may be quite small. The species may be present in many of the same areas both before and after global warming. The American robin, whose large range size is discussed earlier, could be expected to behave this way.

Of the species that have very small ranges, many live in mountainous areas. To date, human activity has had relatively small effects on such species because of the impracticalities of cultivating land in mountains, especially on steep slopes. But it is these species—living on cool mountaintops that are now becoming too warm for them—that have nowhere else to go. Such arguments lead scientists to believe that extinctions caused by rising temperatures will be additional to those caused by land-use changes such as deforestation. Rough calculations suggest that rising temperatures may threaten about a quarter of the species in hot spots.

The small geographic range of Baikal seals (Phoca sibirica)*, endemic to Lake Baikal, southeastern Siberia, Russia, makes them especially vulnerable to extinction.* © Doug Allan/Oxford Scientific Films Ltd.

WHICH SPECIES ARE MOST VULNERABLE TO EXTINCTION?

Before a species becomes extinct, it must first be rare. Some species are naturally rare, while others have that rarity imposed upon them by one or more of the factors discussed earlier. Some species are much more vulnerable than others to becoming dangerously rare, and other species, when rare, are more likely than others of equal rarity to succumb. These and other factors affecting a species' susceptibility to destruction are discussed in this section.

Endemism and Rarity

As previously discussed, a small geographic range makes a species particularly vulnerable to global extinction. Many of the threats to species are geographically restricted, so species with large ranges will survive somewhere even if they are locally extirpated. Species with small ranges do not have this "reserve." Moreover, as also previously mentioned, species with small geographic ranges tend to be rare within them—they have low population densities.

The collections of plants and animals on islands are often rich in endemic species; those species that are restricted to particular islands must inevitably have small ranges. Since 1600 the great majority of known bird extinctions have been on islands. The problem of extinction is not just limited to islands, however, as the previous sections amply describe. In sum, combined endemism and rarity are the factors that confer the greatest risk of extinction.

Body Size and Related Reproductive Characteristics

Generally, the larger the body size of an animal, the longer it lives and the fewer offspring it produces each year. Relatively large animals also tend to have relatively low population densities. Thus, a viable population of, say, elephants occupies considerably more space than an equal-sized population of rabbits. Large predators such as tigers (*Panthera tigris*) have lower population densities than the herbivores on which they feed. A tiger has a home range that may occupy 100 square km (40 square miles), while a rabbit may survive in 1 hectare (0.004 square mile).

Large animals, by virtue of their low population densities, are at increased risk of extinction. Moreover, an animal species that produces few offspring each year and that suffers a major loss in numbers from human activity will need more time to recover than a species with high reproductive rates. Again taking the tiger as an example,

A tiger on a road near Ranthambhor, eastern Rajasthan, India. Larger animals tend to produce fewer offspring, making them more vulnerable to extinction. G. Ziesler/Bruce Coleman Ltd.

a tiger population large enough to avoid all the perils of low population size (including the chance that all offspring will be the same sex or that genetic inbreeding will occur) requires extremely large areas. Although the tiger is not considered an endemic species—it has a very large geographic range, stretching over much of eastern and southern Asia—its rarity in that range means that it is nonetheless vulnerable to extinction from human activity.

Behaviour

At the species level certain behaviours may expose animals and other organisms to increased extinction risk. Some risky behaviours involve the species' relationship with humans, whereas others manifest themselves during mating and dispersal.

Size of Home Range

In the case of lions, wolves, and wild dogs, population size alone is a poor predictor of their local extinction, even when the animals live in protected areas. For these species, conflict with people on the borders of the protected areas is the major cause of mortality. Border areas are population "sinks," areas where the death rate of the population exceeds the birth rate and into which individuals enter through migration from more-central portions of the protected area. This explains, for example, why African wild dogs (*Lycaon pictus*) are under particular threat—more so than other species, such as cheetahs and leopards, that also are afforded protected areas. Wild dogs live in large packs that roam very large areas, in contrast to the two cat species, which are mostly solitary and which have smaller home ranges. The more widely a species moves (i.e., the larger its home range), the more likely it is to move beyond areas where it is protected.

Concentration

Some species have aggregation behaviours that make them vulnerable to disturbance or hunting. For example, bats may congregate in large numbers in particular caves to have their young, making significant portions of their total population especially susceptible when their habitat is disturbed by human visitation or damaged by the cave's commercialization or flooding. The various species of groupers often come together to spawn on a few nights of each year tied to phases of the moon and at traditional mating sites. Fishermen who know these sites and the timing of spawning can devastate large populations of these species by concentrating their efforts during this most vulnerable time in the fish's life cycle.

Low Dispersal

Because small populations are so much more at risk than large ones, individuals of species that can readily disperse can rescue local populations on the verge of extinction. The previously discussed example of the checkerspot butterfly illustrates, first, the rescue of some small populations by the dispersal of individuals from larger "reservoir" populations nearby and, second, the subsequent extinction of the reservoirs as urban development isolated them from still other butterfly populations that could have rescued them. Species that have low dispersal rates are at a disadvantage because it is unlikely that one population can save another.

Mating Systems

Small populations suffer from inbreeding, an inevitable tendency of mating individuals in a small isolated population to be more closely related than they would be in a larger one. When population size is severely reduced, inbreeding may be the final insult that will cause the remaining population to go extinct. The likelihood that this will happen, however, seems related to the social structure of the species involved.

A comparison of two species, a seal and a rhinoceros, serves to explore the issue in more detail. The northern elephant seal (*Mirounga angustirostris*) of the Pacific Coast of North America was thought to have been hunted to extinction in the late 1800s, though it later became apparent that perhaps 20–30 individuals persisted locally for a couple of decades before the population began to recover gradually under protection. The Indian rhinoceros (*Rhinoceros unicornis*) in the early 20th century was reduced to two isolated populations—one numbering between 12 and 100, the other between 60 and 80—before protection allowed it to make a limited recovery. Moreover, not all of the rhinoceros males in the reduced population were likely to have bred. Today the elephant seal is genetically uniform, suggesting that a high degree of inbreeding occurred during the time its population was at a minimum, whereas the rhino has probably lost little of its genetic variability. The population histories of the two species are similar, so why the differences in their genetic variability?

The social structures of elephant seals and rhinos are dramatically different. Each year, the one dominant bull seal that guards the harem is likely to father all of the young. An isolated seal population thus may become genetically uniform relatively quickly because very few males father each generation of young. Rhinos, on the other hand, are largely monogamous, so a group of them will have a greater number of fathers than a comparably sized group of seals.

Insularity

The vulnerability of island species is likely a combination of two factors previously discussed—their endemism and rarity and their ecological naivete, the latter being exemplified by the greater effect of domestic cat introductions on unwary island bird species than on more "streetwise" mainland species. Nevertheless, some island bird species are less likely to be threatened than similar bird species found on continents. The reason lies in the abundance of island species—they are often quite numerous on their islands, for they have fewer competitors than do mainland species.

Human Use

Many species are hunted for meat and other products, including whales and various fish. Less familiar is the

Zebras in the Amboseli National Park, Kenya. A lesser-known trade in the hunting of bushmeat, which basically includes anything that can be hunted, particularly flourishes in Africa. John Lewis Stage—Photo Researchers

widespread trade in bushmeat, which is essentially everything that can be hunted—from mice to chimpanzees and gorillas—and is especially prevalent in West and Central Africa. Yet other species are harvested for body parts, such as tiger bones and rhino horns, which are used in Asian medicines. A wide variety of plants are harvested too, again often for medicinal purposes. Simply put, any species that is used for food, wood, or medicine or as pets or houseplants, that is collected (such as butterflies or invertebrate shells), or that attracts attention for any other reason suffers an increased risk of extinction.

CHAPTER 3
PREVENTING THE LOSS OF BIODIVERSITY

A thorough knowledge of the factors that cause extinction and the vulnerability of different species to them is an essential part of conserving species. In large part, conservation is about removing or reducing those factors and doing so for the most vulnerable species and in the places where species are most vulnerable. Much of the task of conservation professionals is to protect habitats large enough to house viable populations of species, first deciding where the priorities should be and sometimes restoring habitats that already have been destroyed. Local conservation groups often spend time removing introduced species, which can mean physically weeding invasive plants or trapping invasive animals. These activities must be accompanied by efforts to prevent introductions of new threats. Others work to reduce harvesting directly or to reduce the incidental catch of nontarget species. Nonetheless, there are a variety of specific tools that can be applied to different circumstances, as categorized in this chapter and illustrated by case histories. Sometimes, when working with very rare species, scientists may not know the exact causes of threat, which can lead to intense arguments about exactly how to proceed.

SPECIES INTERVENTIONS

Some species become so rare that there are doubts about whether they will be able to survive in the wild. Under such circumstances, the species may be brought into protective custody until areas can be made suitable for their release back into the wild. In addition, inbred populations may benefit from the addition of genes from another population or closely related subspecies. Ensuring that protected species survive is both costly and difficult. Often, the success of a species protection program depends on the willingness of politicians and citizens to enact laws that emphasize wildlife protection and that control exotic species that may have moved into the old habitat areas of the protected species.

Protective Custody

Protective custody is an important tool in plant conservation, where a large number of seeds can be easily stored. In addition, botanic gardens can grow rare plants, protecting the species until such times as they can be planted in the wild. Of a total of about 300,000 described species of flowering plants, Botanic Gardens Conservation International has estimated that about 80,000 species are protected in botanic gardens and a few thousand additional species in other facilities—together about one-third of the total.

For animals, zoos provide an important refuge for some vertebrate species but not for the vast majority of animals, which are invertebrates. Of species protected in zoos, a number of them have later been returned to the wild. To do so, however, substantial problems need to be overcome.

Determining that a species should be brought into captivity and then deciding what to do with the individuals are illustrated by the California condor. The first key decision was whether to bring the birds into protective custody or to manage their small population in the wild. There was no dispute that the condors were once widespread, ranging across the southern and western United States and northern Mexico. As long ago as the turn of the 20th century, however, they were restricted to the mountains of southern California. The widespread practice of setting out poisoned carcasses to kill livestock predators was likely the major cause of their decline. Exactly how many condors still survived was at the core of the debate. Some thought their numbers had been declining constantly, from 150 in the 1950s to 60 in 1970. Others posited constant numbers, and, according to one opinion, if there was no decline, there was need for neither explanation nor intervention—the condors, though rare, should be left alone.

Eventually, photographic surveys completed the catalog of individuals, removing doubts about which individuals were alive, which were dead, and why. Studies of nesting showed that those birds that bred did so with reasonable success. Radio-collared birds showed that the species foraged far beyond the remote areas of its nesting sites, so simply leaving the birds alone and protecting the habitat near the nest would not be sufficient. Birds died from ingesting lead in animals that had been shot or the toxic substances in poison-laced carrion. In the 1980s all the remaining birds were brought into captivity, although not without a lawsuit over the issue.

Because captive populations are almost always small, there is a high risk of inbreeding. Thus, the condors were carefully screened genetically to ensure that as much as possible of the genetic variability of the species was preserved. The captive-breeding program was ultimately a success, and between 12 and 20 fledglings were produced each year after 1991. Earlier production had been lower, likely as a consequence of the birds' inexperience in nesting and rearing. Efforts to teach captive-reared birds through artificial means—some nestlings were fed by puppets resembling condor heads—were much less successful. These birds were particularly inept when released, as real parents teach their young many important things

California condor (Gymnogyps californianus). *Much debate ensued over how best to increase the dwindling numbers of condors.* John Borneman—The National Audubon Society Collection/Photo Researchers

about living in the wild. Eventually, scientists released birds in Arizona, California, and Baja California, and some of these birds have reared young of their own in the wild.

Genetic Intervention

In small populations, inbreeding can cause genetic variability to be lost quite quickly. A simple example is provided by the Y chromosome in humans (and other mammals), which confers maleness and which behaves like human surnames do in large parts of the world. If every human couple had just two children each generation, by chance alone 25 percent of the couples would have two sons, each with one Y chromosome from the father; 25 percent would have two daughters, each with no Y chromosome; and the remaining 50 percent would have one son and one daughter. The surnames of the fathers with two daughters would be lost as they married and had their own children, as indeed would be those fathers' Y chromosomes. In a small breeding population, after just a few generations, every individual would have the last name of the same male ancestor—perhaps a great-great-grandfather—and the same Y chromosome.

Many genetic mutations are deleterious, reducing the individual's chances of survival, but are also recessive, requiring the inheritance of one mutant gene from each parent to manifest their effect. This means that in a large population, their effect will be masked by the overwhelming numerical superiority of the normal dominant gene (allele). For the reasons explained earlier, in a small population chance events cause the loss of genetic variability and so increase the likelihood that individuals will suffer harmful genetic diseases.

Studies on mammals kept in zoos illustrates the harmful effects of inbreeding. In the past, to maintain sufficient productivity, zookeepers often bred animals that proved to be good at producing young. Because of this practice, some breeding pairs quickly came to have the same grandparents and, in some cases, the same parents. The studies showed that such pairs produced young that were much less likely to survive than young from pairs of unrelated individuals.

The practical problem for conservation is whether to place efforts on genetic intervention—bringing in "new blood," that is, individuals and so genes, from the outside—or to concentrate on the factors causing the initial decline.

This issue was at the heart of the management dilemma posed by the Florida panther (*Puma concolor coryi*), a distinct subspecies of puma (*P. concolor*) confined to a small,

Conservation scientists introduced pumas from another area in hopes of increasing the numbers of healthy Florida panthers (Puma concolor coryi). Courtesy, Stuart L. Pimm

isolated, and inbred population in southern Florida. The specific question was whether to introduce pumas from Texas into the Florida population. Florida panthers once had been part of a continuous widespread population. In the 19th century they became isolated in southern Florida. As their numbers declined, the occurrence of genetic defects increased, including sperm and heart defects and undescended testicles in males. Conservation scientists hoped that the introduction of pumas from outside Florida would reverse the genetic damage. This proposal was highly controversial. As with the example of the California condor, some argued that the population was doing well in its limited range and did not need additional animals. Despite concerns, in the mid-1990s eight females from Texas were released, and scientists closely followed the fates of them, of their young, and of the young of cats with only Florida parents. It was found that, although hybrid cats—those with both Florida and Texas parents—do not seem to live longer than pure Florida cats and hybrid females do not produce more kittens than pure Florida females, hybrid kittens survive about twice as well. Since the introduction of the Texas females, numbers of Florida panthers have increased, and hybrid cats were expanding the known range of their habitats.

The Difficulties and Costs of Intervention

Previous discussions have covered the release of individuals into the wild to reestablish populations that have gone extinct or to rescue inbred populations. They also have noted that protecting species from human harvesting and removing invasive species are major techniques in saving biodiversity. These and other species-by-species interventions are often high-profile actions that attract considerable media attention. This does not mean that they are easy or cheap, as the following examples demonstrate.

Protecting Species

For species that are hunted or collected, direct protection may be an essential conservation tool. National laws, such as the Endangered Species Act in the United States, make collecting or killing an endangered species or threatened species illegal. An example of such a protected species in the United States is the country's national bird, the bald eagle (*Haliaeetus leucocephalus*). International laws protect whales of various species, and such agreements as the Convention on International Trade in Endangered Species prohibit commercial trade in designated species. Enforcing these laws and conventions is another matter, however.

The Bald eagle (Haliaeetus leucocephalus) *is just one of many species protected by the Endangered Species Act.* Alan & Sandy Carey/Getty Images

The problems of implementing protection are illustrated by the conservation of the two African species of rhinoceros. The population of the black rhinoceros (*Diceros bicornis*) fell to about 2,400 individuals in 1995, down from a likely number of several hundred thousand at the start of the 20th century, when it ranged over most of southern Africa. The white rhinoceros (*Ceratotherium simum*) historically had a smaller geographic range. Today its northern subspecies occurs only in the Democratic Republic of the Congo, where it is extremely rare. The southern subspecies lives almost entirely in South Africa, Namibia, Zimbabwe, and Kenya. Together, they numbered under 12,000 in 2001, again likely a small fraction of their original numbers.

White rhinoceros (Ceratotherium simum)*, Mkuze Game Reserve, South Africa. The geographic range of white rhinos is quite small.* © Digital Vision/ Getty Images

Although conversion of habitat land to agricultural use and sport hunting caused the earlier rhino declines, the major threat is now poaching—entering reserves where the animals are protected and killing them for their horns, which are in high demand in parts of the world for dagger handles and, in powdered form, as an ingredient of traditional medicines. The costs of protecting rhinos are considerable. WWF has estimated that it costs $1,400 per square km ($3,600 per square mile) per year to detect and deter poaching. Reserves that hold rhinos often cover tens of thousands of square kilometres. The multimillion-dollar budgets required are beyond the means of many African governments.

THE UNITED STATES ENDANGERED SPECIES ACT

In the United States, the U.S. Fish and Wildlife Service (USFWS) of the Department of the Interior and the National Oceanic and Atmospheric Administration (NOAA) of the Department of Commerce are responsible for the conservation and management of fish and wildlife resources and their habitats, including endangered species. The Endangered Species Act of 1973 obligates federal and state governments to protect all life threatened with extinction, and this process is aided by the creation and continued maintenance of an endangered species list, which contains about 1,890 domestic species of endangered or threatened animals and plants. According to the USFWS, the species definition also extends to subspecies or any distinct population segment capable of interbreeding. Consequently, threatened subsets of species may also be singled out for protection. Furthermore, provisions for threatened species—that is, any species expected to become endangered in the future within a substantial portion of its geographic home range—are also included in this law. It also promotes the protection of critical habitats (that is, areas designated as essential to the survival of a given species).

The Endangered Species Act is credited with the protection and recovery of several prominent species within the borders of the United States, such as the bald eagle (*Haliaeetus leucocephalus*), the American alligator (*Alligator mississippiensis*), and the gray wolf (*Canis lupus*).

Removing Invasive Species

Many conservation programs have tackled invasive species. Once again, the message from the two case studies that follow is that, although these programs can be successful, they are often expensive.

As previously described, introduced domestic cats have caused the extinction of many island species. In the 1970s scientists estimated that the cat population on Marion Island, one of the two Prince Edward Islands in the subantarctic Indian Ocean, was killing 450,000 seabirds each year, jeopardizing the birds' survival. The cats, which were estimated to number at least 2,000, were descended from five animals brought to the island in 1949 to deal with a mouse problem at a meteorological station there. Islands are small and isolated, and because cats are predators, they tend to have much smaller populations than their prey. Nonetheless, it took 19 years of an intensive eradication program to remove the last cats.

The second case study is the purple loosestrife (*Lythrum salicaria*), a plant that has overrun thousands of square kilometres of North American wetlands, replacing the naturally diverse vegetation of grasses, sedges, and other wetland plants. It is native to Europe and was introduced into North America in the early 1800s. It now occurs across most of the continental United States and is most prevalent in the northeastern and north-central part of the country and in Canada.

Like other exotic weeds, purple loosestrife can spread rapidly. Control measures have included the use of herbicides and the release of herbivorous insects, including different species of beetles that eat the leaves, roots, and seeds of the plant. The latter strategy, an instance of biological control, is not risk-free. It involves introducing a second nonnative species to control a nonnative species introduced earlier. The effects of the insect introductions have not always been clear, and, as with other examples of biological control, there is a danger that the introduced herbivorous insects will harm native plant species.

Introducing Species

The introductions of Texas pumas into the Florida panther population and of captive-reared condors back into parts of their original habitat were successful, as discussed earlier. So were the introductions to North America of the starling, also previously discussed, and the house sparrow (*Passer domesticus*), which was introduced to New York City from Europe in the 1850s. What is often overlooked, however, is that many other attempts have failed. This in fact is the typical result.

For example, after extensive efforts in the 19th and early 20th centuries to introduce nonnative game birds to the United States, it was found that, even with releases of large numbers of birds, most of the attempts failed. An exception has been the common, or ring-necked, pheasant (*Phasianus colchicus*), native to China and introduced to the United States in the 1890s. This low rate of success has an important implication—even when it is known from hindsight that an individual introduction can succeed, as did the pheasant, most such introductions still will fail. When returning species to the wild that have already gone extinct there, the prognosis is even bleaker. One reason in some cases is that the cause of the extinctions—for instance, the brown tree snake responsible for bird extinctions on Guam—is still present.

INTRODUCED SPECIES

Introduced species, also called alien species or exotic species, are species introduced either accidentally or deliberately by human actions into places beyond its natural geographical range. Familiar examples include the house sparrow, domestic pigeon, and starling, which were all deliberately introduced into North America and other places from their native ranges in Europe. Rats of several species were unintended stowaways on oceanic voyages, probably those of Columbus and certainly those of the Polynesians as they colonized Pacific islands. Infected humans took the human immunodeficiency virus (HIV), which causes AIDS, from Africa to the rest of the world. In much the same way, early European colonists of the New World and the Pacific took other sexually transmitted microbes, common cold viruses, smallpox viruses, and a variety of other disease-causing organisms to populations with no resistance to them. For some species, the introduction was deliberate but the spread of the species and its consequences were unexpected. For example, kudzu, a vine introduced from Japan to the southeastern United States to stabilize steep banks of soil created by road construction, has spread up and over large parts of the native forest, choking it.

Introduced species sometimes cause massive economic harm. For example, introduced cacti in many parts of the world reduce productive rangeland to impenetrable, thorny bushland. In addition, by crowding out a region's competitively inferior indigenous species, introduced species are a major factor in the extinction of species and consequent loss of biological diversity.

HABITAT PROTECTION

Because the loss of habitat is the primary reason that species are lost both locally and globally, protecting more habitat emerges as the most important priority for conservation. This simple idea raises difficult questions. Which habitats should be protected? And because it seems unlikely that all habitats can be protected, which ones should receive priority?

If reserves were judiciously placed over the identified hot spots of biodiversity, the special places where vulnerable species are concentrated, a large fraction of species might be saved. Presently, the allocation of reserves around the world is poor. Reserves larger than 100,000 square km (40,000 square miles) are generally in high mountains, tundras, and the driest deserts, areas that are not particularly species-rich. On the other hand, hot spots such as Madagascar and the Philippines protect less than 2 percent of their land.

The same kinds of questions hold on smaller scales, as illustrated by a study reported in the late 1990s. The Agulhas Plain at the southern tip of Africa is one of the world's "hottest" spots for concentrations of vulnerable plant species. An area only some 1,500 square km (600 square miles) in size was found to house 1,751 plant species, 99 of them endemic. Whereas most of the state forests and private nature reserves in the area are coastal, most of the hot spot's endemic plants live inland. Given that new reserves must be created if these plants are to survive, where should they be situated to encompass the maximum number of species at minimum cost?

Fortunately, the data available to make these decisions included a knowledge of the distribution of plant species over the Agulhas Plain in fairly good detail—the kind of information not likely to be available in most hot spots. This allowed the plain's plant-species composition to be divided into a grid of cells, each 3 × 3 km (1.8 × 1.8 miles) on a side. Computer algorithms (systematic problem-solving methods) were then used to select sets of cells from the grid according to their complementary species composition—that is, the aim was to encompass as many species or as many endemics as possible in as small an area (as few grid cells) as possible.

Naively applied, these algorithms will not give useful results. For example, the sites they select may not be available for reserves. Also, the choice of too small a cell size can lead to the selection of protected areas containing populations so small and widely scattered that they would be unlikely to persist. This is fittingly dubbed the "Noah's Ark effect," because the ark held only two individuals of each species for a short time. Reserves need to be large enough to support species indefinitely. The choice of a cell size of 3 × 3 km (1.8 × 1.8 miles) is politically feasible because reserves of this size already have been established in the region and are probably ecologically sensible for many plant species. Other factors had to be included in the final selection of cells. Some areas are unsuitable for various reasons—for instance, some are overrun by alien weeds, while others are mostly in urban areas or croplands. In contrast, other areas are particularly desirable—for example, they may be adjacent to existing reserves, and it is easier to expand such reserves than to create new ones. The results of this study thus provided advice for establishing reserves that combined ecological information on species distribution with practical and political considerations.

Saving the most species for the least money likewise was the consideration that motivated another study published in the late 1990s, of which counties in the United States should be conservation priorities. An earlier study that attempted to locate sites for new reserves in the United States had equated efficiency with the minimum number of counties needed to achieve a given coverage of endangered species. That approach would have been sensible if land were much the same price everywhere. Unfortunately, the study's targets had included counties encompassing San Diego, Santa Cruz, and San Francisco in California, Honolulu in Hawaii, and certain counties in Florida, all of which are among the highest-priced land in the country.

The later study asked how many species could be protected for a given total cost. It found that considerable savings in cost per species accrue from selecting larger, more-complementary areas and lower total costs and that, as a consequence of this approach, the places identified for protection were often quite different from those recommended in the earlier study.

HABITAT CONNECTIONS

Habitats that are not completely destroyed may be fragmented to the degree that individual fragments are too small to hold viable populations of many species—they may suffer from inbreeding or the increased demographic risks previously discussed. Yet in total the fragments may actually be of sufficient area to support these species. An obvious conservation intervention is to find ways of connecting fragments by corridors. These corridors can be created from currently unprotected land between existing reserves or by restoring land between existing habitat fragments.

Many contemporary efforts to create corridors occur on a small scale. They can be as simple as hedgerows that connect woodlots, a strategy that likely works for some small species. Other efforts are far more ambitious. One of the earliest large efforts involved a plan to connect various parks and other protected areas in Florida by corridors of land that would have to be purchased or otherwise protected from development. From this initial effort, a conservation group, the Wildlands Project, has developed an extensive set of plans for many areas in North America intended to set priorities for the acquisition of land for a mosaic of corridors that would eventually link together large parts of the continent.

A regional example of corridor creation is in Costa Rica. Situated in the Caribbean lowlands, La Selva Biological

Station is one of the major centres for research on tropical forests. Occupying an area of about 16 square km (6 square miles), the station is bordered on the south by Braulio Carrillo National Park, a much larger area of forest covering 460 square km (180 square miles). The national park extends to La Selva through a forest corridor that descends in elevation from nearly 3,000 metres (nearly 10,000 feet) at Barva Volcano down to 35 metres (115 feet) above sea level at La Selva. The corridor had been threatened by agricultural development until conservation groups, realizing that La Selva would become an isolated forest "island," purchased the corridor to protect it.

Habitat Management

Once protected, areas must often be managed to maintain the threatened species within them. Management may involve the removal of alien species, as previously discussed. It can also involve restoring natural ecological processes to the area. Original fire and flooding regimes are examples of such processes, and they are often controversial because human actions can alter them significantly.

Fire Control

Despite the often valid reasons for suppressing wildfires, the practice can change vegetation dramatically and sometimes harm species in the process. As previously noted, human activities have changed fire regimes across large areas of the planet, including some biodiversity hot spots. Getting the fire regimes right can be essential for conserving species.

An example of a species for which the control of fire regimes has proven both possible and essential is Kirtland's warbler (*Dendroica kirtlandii*). This endangered species nests only in the Upper Peninsula of Michigan, an exceptional case of a bird species with a tiny geographic range well outside the tropics. The bird places its nest in grasses and shrubs below living branches of jack pines (*Pinus banksiana*) that are between 5 and 20 years old. The region's natural wildfires originally maintained a sufficient area of young jack pines. As elsewhere, modern practices suppressed fires, and the habitat declined. Active management of fires to ensure that there are always jack pines of the right age has steadily increased the population of warblers since the early 1990s, when it had included only about 200 singing males. By the early years of the 21st century, there were about 1,300 singing males.

Flood Control

In much the same way that human actions suppress fire regimes, they also control water levels, and the resulting changes can have important consequences for endangered species. An example of a species so affected is the Cape Sable seaside sparrow (*Ammodramus maritimus mirabilis*) found in the Florida Everglades. The Everglades once stretched from Lake Okeechobee in the north to Florida Bay in the south. Water flowed slowly over a wide area, and its levels varied seasonally: summer rains caused the levels to rise in late summer and early fall; then the dry season dropped the levels to their lowest in late May. Under this natural regime, some prairies were continuously flooded for many years, drying out in only the driest years, whereas others were flooded for only a few months each year. It is in these drier prairies that the sparrows nest from about the middle of March until the water floods their nests in the summer.

Water-management actions have diverted the flow of water to the west of its natural path, making the western part of the Everglades unnaturally wet in some years during the bird's nesting season. Unnatural flooding during the four years that followed 1992 reduced the bird population in the west to less than 10 percent of its 1992 level. In the east the prairies have become unnaturally dry and so have become subject to an increased number of fires, which has jeopardized the sparrow population there. It is clear that rescuing the Cape Sable seaside sparrow requires the restoration of the natural water levels and flows within the Everglades and the consequent return to natural fire regimes.

HABITAT RESTORATION

Once a habitat has been destroyed, the only remaining conservation tool is to restore it. The problems involved may be formidable, and they must include actions for dealing with what caused the destruction. Restorations are massive ecological experiments. As such, they are likely to meet with different degrees of success in different places. Restoration of the Everglades, for example, requires restoring the natural patterns of water flow to thousands of square kilometres of southern Florida.

A case history of habitat restoration comes from the Midwestern United States. In Illinois, natural ecosystems cover less than 0.1 percent of the state, so restoration is almost the only conservation tool available. North Branch is a 20-km (12-mile) strip of land running northward from Chicago along the north branch of the Chicago River. Early in the 20th century, it was protected from building but later abandoned. Beginning in the 1970s, a group of volunteers first cleared out introduced European buckthorn (*Rhamnus cathartica*), removed abandoned cars, and planted seeds that they had gathered from combing the tiny surviving remnants of original prairies in places such as along railways and in cemeteries. The initial effort to replant the prairie species failed, however, when animals ate the small growing plants. The solution was to restore the natural fire regime, although controlled burns—setting fires safely near homes—posed a difficult technical challenge. But once it was accomplished, the results were immediate and dramatic. The original prairie plants flourished and the weeds retreated, although with an important exception. Under the native trees grew non-native thistles and dandelions.

The original habitats locally called barrens constituted a visually striking and ecologically special habitat. Restoring them was a particular challenge, and the main conservation problem was finding the right mix of species. One recommendation was to use remnant barrens as models, but the North Branch volunteers rejected them as being too degraded. In the early 20th century, naturalists had speculated that barrens were special because they lacked some characteristic prairie species and, at the same time, had their own distinctive species. The volunteers examined habitat descriptions in old treatises of local plants looking for such species, many of which certainly would now be scarce. The key discovery was a list of barrens plants published by a country doctor in 1846. The scientific names had changed in 150 years, but, by tracking them through the local literature, the volunteers found many of their putative barrens-specific plants on the doctor's list. In possession of this vital information, the group succeeded in restoring barrens sites that by 1991 held 136 native plants, including whole patches consisting of species that, until restoration, had been locally rare.

The preceding example illustrates two rules of ecological restoration. The first is that one must be able to save all the component species of the original ecological community. The second, which is illustrated by the doctor's list, is that one needs to know which species belong and which do not. Without the right species mix, restorers must constantly weed and reseed.

PUTTING A PRICE ON CONSERVATION

Estimating the economic value of biodiversity is controversial. The value of whales, for example, once would have been just the price that traders would have paid for their parts, mostly oil and baleen. Today whales have an obvious economic value that comes from what tourists spend to watch them from boats—vessels that often now sail from the same ports that once supported whaling. Less obvious as an economic factor are the deeply held beliefs that hunting whales is cruel or that whales have a right to exist. Many people throughout the world object to whaling even though they may rarely or never have seen whales. Economists incorporate these ideas into their calculations of value by asking how much the public would pay to protect whales.

A relevant real-world example is found in the aftermath of the oil spill from the tanker *Exxon Valdez*, which ran aground in Prince William Sound, Alaska, in 1989. When the spill polluted the spectacular scenery of Alaska's coast, many Americans felt that they had suffered a personal loss. Exxon was forced to pay for cleaning up the mess and was fined not only for harming people's livelihoods but also for incurring that loss.

Whereas the propriety of decisions that put monetary values on moral issues is contentious, in 1997 a team of ecologists and economists attempted to do just that on a grand scale when they reported their calculations of how much the diverse services provided by Earth's ecosystems are worth. Their approach was to group these services into 17 categories and to classify the planet's surface into 16 ecosystem types. Many combinations of services and ecosystem types were not estimated. Mountains, the Arctic, and deserts certainly have cultural values and devoted ecotourists, but these were absent, as were urban areas with city parks. Nonetheless, the total value of $33 trillion per year at which the team arrived is roughly twice the value of the global economy.

A challenge for today's society is to comprehend the economic values of Earth's ecosystems, for often they are taken for granted. One effort to do this is found in the steps taken by New York City to improve its water supply by protecting the watersheds that supply that water. The cost of this protection was small compared with the alternative of greatly expanding the city's system of water-treatment plants. Another example, aimed at slowing the rate of deforestation, can be seen in the Costa Rican government's willingness to pay for protecting forest watersheds. A global example comprises efforts to reduce carbon emissions worldwide by rewarding those who protect existing forests or restore forests on deforested land.

Governments and international agencies often encourage the destruction of biodiversity by providing financial subsidies in the interests of economic growth. For example, many countries massively support their national fishing fleets by subsidizing the price of fish. The Soviet Union was such a country, but on its breakup the Russian government withdrew the subsidies, with the result that much of the fishing fleet has rusted in harbour. A study conducted in the late 20th century by the Food and Agriculture Organization of the United Nations estimated that the total annual value of the world fish catch was about $70 billion. Yet the cost of catching these fish was $92 billion, reflecting a loss of $22 billion. The cost included routine maintenance, fuel, insurance, supplies, and labour. The magnitude of the loss was certainly an underestimate, for

EXXON VALDEZ OIL SPILL

This massive oil spill occurred on March 24, 1989, in Prince William Sound, an inlet in the Gulf of Alaska, Alaska, U.S. The incident happened after an Exxon Corporation tanker, the *Exxon Valdez*, ran aground on Bligh Reef during a voyage from Valdez, Alaska, to California. Delayed efforts to contain the spill and naturally strong winds and waves dispersed nearly 11,000,000 gallons (41,640 kilolitres) of North Slope crude oil across the sound. The spill eventually polluted 1,300 miles (2,092 kilometres) of indented shoreline, as well as adjacent waters, as far south as the southern end of Shelikof Strait between Kodiak Island and the Alaska Peninsula. Alaska Sen. Ted Stevens emerged as a strong proponent of securing federal funds to pay for the damage. Thousands of workers and volunteers helped to clean up after the oil spill, and Exxon provided $2.1 billion in funding. Despite these cleanup efforts, the spill exterminated much native wildlife, including salmon, herring, sea otters, bald eagles, and killer whales.

The National Transportation Safety Board (NTSB) eventually assigned most of the blame for the oil spill to Exxon, citing its incompetent and overworked crew. The board also faulted the U.S. Coast Guard for an inadequate system of traffic regulation. After evidence suggested that Joseph J. Hazelwood, the ship's captain, had been drinking before the accident, Exxon terminated his employment. The *Exxon Valdez* itself was repaired and recommissioned but was legally prohibited from ever reentering Prince William Sound.

any private company would also include in its total cost the annual costs of paying off the loans for the fishing boats and other facilities. Those loans, the study concluded, would account for another $32 billion, making the annual deficit $54 billion per year.

Agricultural subsides worldwide are even larger—a global review of all subsidies estimates that they total trillions of dollars per year. The review suggests that a powerful strategy for conservation would be to examine whether these subsidies are sensible given concerns about the environment and the need to be fiscally responsible.

CHAPTER 4
ECOLOGY

Ecology, which is also called bioecology, bionomics, or environmental biology, is the study of the relationships between organisms and their environment. Some of the most pressing problems in human affairs—expanding populations, food scarcities, environmental pollution including global warming, extinctions of plant and animal species, and all the attendant sociological and political problems—are to a great degree ecological.

The word *ecology* was coined by the German zoologist Ernst Haeckel, who applied the term *oekologie* to the "relation of the animal both to its organic as well as its inorganic environment." The word comes from the Greek *oikos*, meaning "household," "home," or "place to live." Thus, ecology deals with the organism and its environment. The concept of environment includes both other organisms and physical surroundings. It involves relationships between individuals within a population and between individuals of different populations. These interactions between individuals, between populations, and between organisms and their environment form ecological systems, or ecosystems. Ecology has been defined variously as "the study of the interrelationships of organisms with their environment and each other," as "the economy of nature," and as "the biology of ecosystems."

HISTORICAL BACKGROUND

Ecology had no firm beginnings. It evolved from the natural history of the ancient Greeks, particularly Theophrastus, a friend and associate of Aristotle. Theophrastus first described the interrelationships between organisms and

Ernst Haeckel, c. *1870. Haeckel originally coined the term* ecology. The Bettmann Archive

between organisms and their nonliving environment. Later foundations for modern ecology were laid in the early work of plant and animal physiologists.

In the early and mid-1900s two groups of botanists, one in Europe and the other in the United States, studied plant communities from two different points of view. The European botanists concerned themselves with the study of the composition, structure, and distribution of plant communities. The American botanists studied the development of plant communities, or succession. Both plant and animal ecology developed separately until American biologists emphasized the interrelation of both plant and animal communities as a biotic whole.

During the same period, interest in population dynamics developed. The study of population dynamics received special impetus in the early 19th century, after the English economist Thomas Malthus called attention to the conflict between expanding populations and the capability of Earth to supply food. In the 1920s the American zoologist Raymond Pearl, the American chemist and statistician Alfred J. Lotka, and the Italian mathematician Vito Volterra developed mathematical foundations for the study of populations, and these studies led to experiments on the interaction of predators and prey, competitive relationships between species, and the regulation of populations. Investigations of the influence of behaviour on populations were stimulated by the recognition in 1920 of territoriality in nesting birds. Concepts of instinctive and aggressive behaviour were developed by the Austrian zoologist Konrad Lorenz and the Dutch-born British zoologist Nikolaas Tinbergen, and the role of social behaviour in the regulation of populations was explored by the British zoologist Vero Wynne-Edwards.

While some ecologists were studying the dynamics of communities and populations, others were concerned with energy budgets. In 1920 August Thienemann, a German freshwater biologist, introduced the concept of trophic, or feeding, levels, by which the energy of food is transferred through a series of organisms, from green plants (the producers) up to several levels of animals (the consumers). An English animal ecologist, Charles Elton (1927), further developed this approach with the concept of ecological niches and pyramids of numbers. In the 1930s, American freshwater biologists Edward Birge and Chancey Juday, in measuring the energy budgets of lakes, developed the idea of primary productivity, the rate at which food energy is generated, or fixed, by photosynthesis. In 1942 Raymond L. Lindeman of the United States developed the trophic-dynamic concept of ecology, which details the flow of energy through the ecosystem. Quantified field studies of energy flow through ecosystems were further developed by the brothers Eugene Odum and Howard Odum of the United States. Similar early work on the cycling of nutrients was done by English forester J.D. Ovington.

The study of both energy flow and nutrient cycling was stimulated by the development of new materials and techniques—radioisotope tracers, microcalorimetry, computer science, and applied mathematics—that enabled ecologists to label, track, and measure the movement of particular nutrients and energy through ecosystems. These modern methods encouraged a new stage in the development of ecology—systems ecology, which is concerned with the structure and function of ecosystems.

AREAS OF STUDY

Ecology is necessarily the union of many areas of study because its definition is so all-encompassing. There are many kinds of relationships between organisms and their environment. By *organisms* one might mean single individuals, groups of individuals, all the members of one species, the sum of many species, or the total mass of species (biomass) in an ecosystem. And the term *environment* includes not only physical and chemical features but also the biological environment, which involves yet more organisms.

In practice, ecology is composed of broadly overlapping approaches and further divided by the groups of species to be studied. There are many, for example, who specialize in the field of "bird behavioral ecology." The main approaches fall into the following classes.

Evolutionary ecology examines the environmental factors that drive species adaptation. Studies of the evolution of species might seek to answer the question of how populations have changed genetically over several generations but might not necessarily attempt to learn what the underlying mechanisms might be. Evolutionary ecology seeks those mechanisms. Thus, in the well-known example of the peppered moth, the populations in the industrialized English Midlands changed over generations from having wings coloured largely grayish white, peppered with black spots, to wings that were mostly blackish. The ecological mechanism involved predation—birds readily detected the light-coloured moths against the background of the tree trunks that industrial pollution had darkened, whereas the dark-coloured moths remained generally undetected.

Evolutionary ecology also examines broader issues, such as the observations that plants in arid environments often have no leaves or else very small ones or that some species of birds have helpers at the nest—individuals that raise young other than their own. A critical question for the subject is whether a set of adaptations arose once and has simply been retained by all species descended from a common ancestor having those adaptations or whether the adaptations evolved repeatedly because of the same environmental factors. In the case of plants that live in arid environments, cacti from the New World and euphorbia from the Old World can look strikingly similar even though they are in unrelated plant families.

Physiological ecology asks how organisms survive in their environments. There is often an emphasis on extreme conditions, such as very cold or very hot environments or aquatic environments with unusually high salt concentrations. Examples of the questions it may explore are: How do some animals flourish in the driest deserts, where temperatures are often high and freestanding water is never available? How do bacteria survive in hot springs, such as those in Yellowstone National Park in the western United States, that would cook most species? How do nematodes live in the soils of dry valleys in Antarctica? Physiological ecology looks at the special mechanisms that the individuals of a species use to function and at the limits on species imposed by the environment.

Behavioral ecology examines the ecological factors that drive behavioral adaptations. The subject considers how individuals find their food and avoid their enemies. For example, why do some birds migrate while others are resident? Why do some animals, such as lions, live in groups while others, such as tigers, are largely solitary?

Population ecology, or autecology, examines single species. One immediate question that the subject addresses is why some species are rare while others are abundant. Interactions with other species may supply some of the answers. For example, enemies of a species can restrict its numbers, and those enemies include predators, disease organisms, and competitors (i.e., other species). Consequently, population ecology shares an indefinite boundary with community ecology, a subject that examines the interactions between several to many species.

Species abundances vary both from year to year and across the species' geographic range. Population ecology asks what causes abundances to fluctuate. Why, for example, do numbers of some species, typically birds and mammals, change perhaps threefold or fourfold over a decade or so, while numbers of other species, typically insects, vary tenfold to a hundredfold from one year to the next? Another key question is what limits abundance, for, without limits, species numbers would grow exponentially.

Biogeography is the study of the geographical distribution of organisms, and it asks questions that parallel those of population ecology. Some species have tiny geographical ranges, being restricted to perhaps only a few square kilometres, while other species have ranges that cover a continent. Some species have more-or-less fixed geographical ranges, while others fluctuate, and still others are on the increase. If a species that is spreading is an agricultural pest, a disease organism, or a species that carries a disease, understanding the reasons for the increasing range may be a matter of considerable economic importance. Biogeography also considers the ranges of many species, asking why, for example, species with small geographic ranges are often found in special places that house many such species rather than scattered randomly about the planet.

Community ecology, or synecology, considers the ecology of communities, the set of species found in a particular place. Because the complete set of species for a particular place is usually unknown, community ecology often focuses on subsets of organisms, asking questions, for example, about plant communities or insect communities. A fundamental question deals with the size of the "set of species"—that is, what ecological factors determine how many species are present in an area. There are many large-scale patterns. For example, more species are present in larger areas than smaller ones, more on continents than on islands (especially remote ones), and more in the tropics than in the Arctic. There are many hypotheses for each pattern. Ecological factors also cause the diversity of species to vary over smaller scales. For example, though predators may be harmful to individual species, the presence of a predator may actually increase the number of species present in a community by limiting the numbers of a particularly successful competitor that otherwise might monopolize all the available space or resources.

The preceding questions are generally applied to species at the same trophic level—say, the plants in a community, or the insects that feed on the plants there, or the birds that feed on the insects there. Yet a different set of questions in community ecology involves how many trophic levels there are in a particular place and what factors limit that number.

Conservation biology seeks to understand what factors predispose species to extinction and what humans can do about preventing extinction. Species in danger of extinction are often those with the smallest geographic ranges or the smallest population sizes, but other ecological factors are also involved.

Ecosystem ecology examines large-scale ecological issues, ones that often are framed in terms not of species but rather of measures such as biomass, energy flow, and nutrient cycling. Questions include how much carbon is absorbed from the atmosphere by terrestrial plants and marine phytoplankton during photosynthesis and how much of that is consumed by herbivores, the herbivores' predators, and so on up the food chain. Carbon is the basis of life, so these questions may be framed in terms of energy. How much food one has to eat each day, for instance, can be measured in terms of its dry weight or its calorie content. The same applies to measures of production for all the plants in an ecosystem or for different trophic levels of an ecosystem. A basic question in ecosystem ecology is how much production there is and what the factors are that affect it. Not surprisingly, warm, wet places such as rainforests produce more than extremely cold or dry places, but other factors are important. Nutrients are essential and may be in limited supply. The availability of phosphorus and nitrogen often determines productivity—it is the reason these substances are added to lawns and crops—and their availability is particularly important in aquatic systems. Conversely, nutrients can represent too much of a good thing. Human activity has modified global ecosystems in ways that are increasing atmospheric carbon dioxide, a carbon source but also a greenhouse gas, and causing excessive runoff of fertilizers into rivers and then into the ocean, where it kills the species that live there.

THE CARBON CYCLE

In biology the carbon cycle is the circulation of carbon in various forms through nature. Carbon is a constituent of all organic compounds, many of which are essential to life on Earth. The source of the carbon found in living matter is carbon dioxide (CO_2) in the air or dissolved in water. Algae and terrestrial green plants (producers) are the chief agents of carbon dioxide fixation through the process of photosynthesis, through which carbon dioxide and water are converted into simple carbohydrates. These compounds are used by the producers to carry on metabolism, the excess being stored as fats and polysaccharides. The stored products are then eaten by consumer animals, from protozoans to man, which convert them into other forms. All animals return CO_2 directly to the atmosphere as a by-product of their respiration. The carbon present in animal wastes and in the bodies of all organisms is released as CO_2 by decay, or decomposer, organisms (chiefly bacteria and fungi) in a series of microbial transformations.

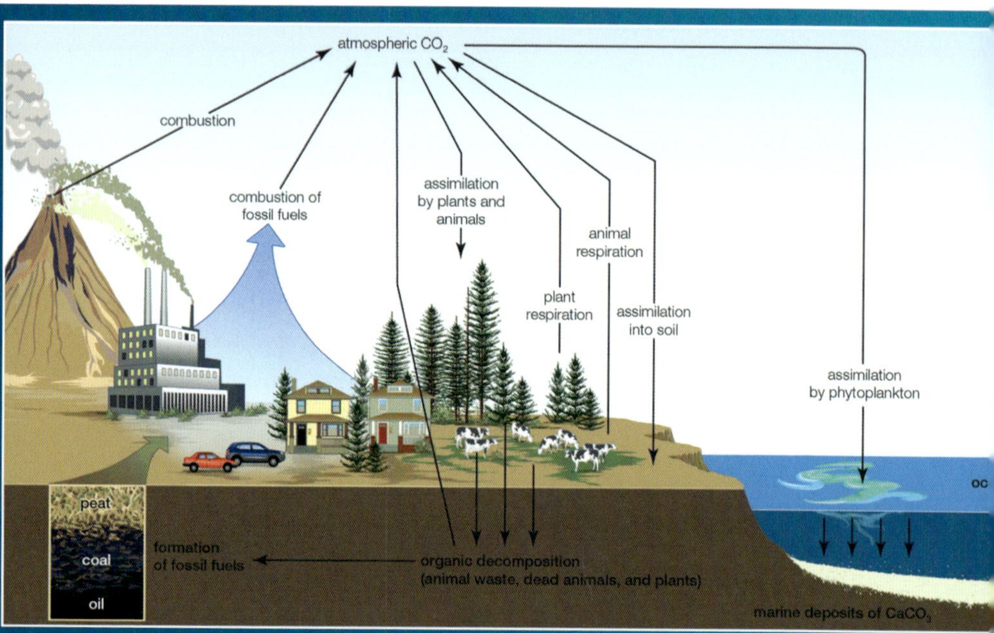

The generalized carbon cycle. Encyclopædia Britannica, Inc.

Part of the organic carbon—the remains of organisms—has accumulated in the Earth's crust as fossil fuels (e.g., coal, gas, and petroleum), limestone, and coral. The carbon of fossil fuels, removed from the cycle in prehistoric time, is now being released in vast amounts as CO_2 through industrial and agricultural processes, much of it quickly passing into the oceans and there being "fixed" as carbonates. If oxygen is scarce (as in sewage, marshes, and swamps) some carbon is released as methane gas.

METHODS IN ECOLOGY

Because ecologists work with living systems possessing numerous variables, the scientific techniques used by physicists, chemists, mathematicians, and engineers require modification for use in ecology. Moreover, the techniques are not as easily applied in ecology, nor are the results as precise as those obtained in other sciences. It is relatively simple, for example, for a physicist to measure gain and loss of heat from metals or other inanimate objects, which possess certain constants of conductivity, expansion, surface features, and the like. To determine the heat exchange between an animal and its environment, however, a physiological ecologist is confronted with an array of almost unquantifiable variables and with the formidable task of gathering the numerous data and analyzing them. Ecological measurements may never be as precise or subject to the same ease of analysis as measurements in physics, chemistry, or certain quantifiable areas of biology.

In spite of these problems, various aspects of the environment can be determined by physical and chemical means, ranging from simple chemical identifications and physical measurements to the use of sophisticated mechanical apparatus. The development of biostatistics (statistics applied to the analysis of biological data), the elaboration of proper experimental design, and improved sampling methods now permit a quantified statistical approach to the study of ecology. Because of the extreme difficulties of controlling environmental variables in the field, studies involving the use of experimental design are largely confined to the laboratory and to controlled field experiments designed to test the effects of only one variable or several variables. The use of statistical procedures and computer models based on data obtained from the field provide insights into population interactions and ecosystem functions. Mathematical programming models are becoming increasingly important in applied ecology, especially in the management of natural resources and agricultural problems having an ecological basis.

Controlled environmental chambers enable experimenters to maintain plants and animals under known conditions of light, temperature, humidity, and day length so that the effects of each variable (or combination of variables) on the organism can be studied. Biotelemetry and other electronic tracking equipment, which allow the movements and behaviour of free-ranging organisms to be followed remotely, can provide rapid sampling of populations. Radioisotopes are used for tracing the pathways of nutrients through ecosystems, for determining the time and extent of transfer of energy and nutrients through the different components of the ecosystem, and for the determination of food chains. The use of laboratory microcosms—aquatic and soil micro-ecosystems, consisting of biotic and nonbiotic material from natural ecosystems, held under conditions similar to those found in the field—are useful in determining rates of nutrient cycling, ecosystem development, and other functional aspects of ecosystems. Microcosms enable the ecologist to duplicate experiments and to perform experimental manipulation on them.

POPULATION ECOLOGY

Population ecology is the study of the processes that affect the distribution and abundance of animal and plant populations. A population is a subset of individuals of one species that occupies a particular geographic area and, in sexually reproducing species, interbreeds. The geographic boundaries of a population are easy to establish for some species but more difficult for others. For example, plants or animals occupying islands have a geographic range defined by the perimeter of the island. In contrast, some species are dispersed across vast expanses, and the boundaries of local populations are more difficult to determine. A continuum exists from closed populations that are geographically isolated from other populations of the same species to open populations that show varying degrees of connectedness.

Genetic Variation Within Local Populations

In sexually reproducing species, each local population contains a distinct combination of genes. As a result, a species is a collection of populations that differ genetically from one another to a greater or lesser degree. These genetic differences manifest themselves as differences among populations in morphology, physiology, behaviour, and life histories. In other words, genetic characteristics (genotype) affect expressed characteristics (phenotype). Natural selection initially operates on a phenotypic level, favouring or discriminating against expressed characteristics. The gene pool (total aggregate of genes in a population at a certain time) is affected as organisms with phenotypes that are compatible with the environment are more likely to survive for longer periods, during which time they can reproduce more often and pass on more of their genes.

The amount of genetic variation within local populations varies tremendously, and much of modern conservation biology is concerned with the maintenance of genetic variation within populations of plants and animals. Some small, isolated populations of asexual species often have little genetic variation among individuals, whereas large, sexual populations often have great variation. Two major factors are responsible for this variety: mode of reproduction and population size.

The Effects of Sexual and Asexual Reproduction

In sexual populations, genes are recombined in each generation, resulting in new genotypes. Offspring in most sexual species inherit half their genes from their mother and half from their father, and their genetic makeup is therefore different from either parent or any other individual in the population. New, favourable mutations that initially appear in separate individuals can be recombined in many ways over time within a sexual population. In contrast, the offspring of an asexual individual are genetically identical to their parent. The only source of new gene combinations in asexual populations is mutation.

Asexual populations accumulate genetic variation only at the rate at which their genes mutate. Favourable mutations arising in different asexual individuals have no way of recombining and eventually appearing together in any one individual, as they do in sexual populations. A much wider range of favourable gene combinations, therefore, is likely to be found in sexual than asexual populations.

The Effects of Population Size

Over long periods of time, genetic variation is more easily sustained in large populations than in small populations. Through the effects of random genetic drift, a genetic trait can be lost from a small population relatively quickly. For example, many populations have two or more forms of a gene, which are called alleles. Depending on which allele an individual has inherited, a certain phenotype will be produced. If populations remain small for many generations, they may lose all but one form of each gene by chance alone.

This loss of alleles happens from sampling error. As individuals mate, they exchange genes. Imagine that initially half of the population has one form of a particular gene, and the other half of the population has another form of the gene. By chance, in a small population the exchange of genes could result in all individuals of the next generation having the same allele. The only way for this population to contain a variation of this gene again is through mutation of the gene or immigration of individuals from another population.

Minimizing the loss of genetic variation in small populations is one of the major problems faced by conservation biologists. Environments are constantly changing, and natural selection continually sorts through the genetic variation found within each population, favouring those individuals with phenotypes best suited for the current environment. Without the genetic variation that allows populations to respond evolutionarily to changes in the physical environment, diseases, predators, and competitors, populations risk extinction.

Population Density and Growth

The development of an effective species protection or recovery program requires that wildlife managers know quite a bit about the species under their supervision. The construction of life histories and life tables are useful for understanding the timing of significant life events, such as reproduction and dispersal. These tools help officials design programs that maximize the reproductive capacity of a species under threat.

Life Histories and the Structure of Populations

A life history is the sequence and timing of events that occur between birth and death. Populations from different parts of the geographic range that a species inhabits may exhibit marked variations in their life histories. The patterns of variation seen within and among populations are referred to as the structure of populations. These variations include breeding frequency, age at which reproduction begins, the number of times an individual reproduces during its lifetime, the number of offspring produced at each reproductive episode (clutch size), the ratio of male to female offspring produced, and whether reproduction is sexual or asexual. These differences in life history characteristics can have profound effects on the dynamics, ecology, and evolution of populations.

Of the many differences in life history that occur among populations, age at the time of first reproduction is one of the most important for understanding the dynamics and evolution of a population. All else being equal, natural selection will favour individuals that reproduce earlier than other individuals within a population, because by reproducing earlier an individual's genes enter the gene pool sooner than those of other individuals that were born at the same time but have not reproduced. The genes of the early reproducers then begin to spread throughout the population. Individuals whose genetic makeup allows them to reproduce earlier in life will come to dominate a population if there is no counterbalancing advantage to those individuals that delay reproduction until later in life.

Not all populations, however, are made up of individuals that reproduce very early in life. In the course of a lifetime, an individual must devote energy and resources to physiological demands other than reproduction. To reproduce successfully, a plant first may have to grow to a certain height and outcompete its neighbours, and an animal may have to devote energy to growth so that it can reach a size at which it can fend off predators and successfully compete for mates. In many populations, individuals that delay reproduction have a better chance of surviving and leaving offspring than those that attempt to reproduce early. The opposing demands of growth, defense, and reproduction are balanced within the constraints of different environments to produce populations that have a diverse range of life-history strategies.

EFFECTIVE POPULATION SIZE

In genetics the size of a breeding population is known as the effective population size. Although the total population of a species may be many times larger than the breeding population, nonbreeding individuals do not directly contribute to the numerical success of the population over the long term, and thus they are far less important than breeding individuals in population studies and conservation management. The number of individuals that breed in a population is determined by the number of parents, the average number of children per family, and the extent to which family size varies from the average. The determination of the effective population size of a breeding population is necessary for studies of population growth rates and of gene flow.

Populations often can be divided into one of two extreme types, based on their life-history strategy. Some populations, called *r*-selected, are considered opportunistic because their reproductive behaviour involves a high intrinsic rate of growth (*r*)—individuals give birth once at an early age to many offspring. Populations that exhibit this strategy often have been shaped by an extremely variable and uncertain environment. Because mortality occurs randomly in this setting, quantity of progeny rather than quality of care serves the species better. In another strategy, called *K*-selected, populations tend to remain near the carrying capacity (*K*), the maximum number of individuals that the environment can sustain. Individuals in a *K*-selected population give birth at a later age to fewer offspring. This equilibrial life history is exhibited in more stable environments where reproductive success depends more on the fitness of the offspring rather than on their numbers.

Life Tables and the Rate of Population Growth

Differences in life-history strategies greatly affect population dynamics. As stated earlier, populations in which individuals reproduce at an early age have the potential to grow much faster than populations in which individuals reproduce later. The effect of the age of first reproduction on population growth can be seen in the life tables for a particular species. Life tables were originally developed by insurance companies to provide a means of determining how long a person of a particular age could be expected to live. They are used not only by demographers of human populations but also by plant, animal, and microbial ecologists to make projections about the life expectancies of nonhuman populations. The number of individuals in a closed population (a population in which neither immigration nor emigration occurs) is governed by the rates of birth (natality), growth, reproduction, and death (mortality). Life tables are designed to evaluate how these rates influence the overall growth rate of a population.

Survivorship Curves

Life tables follow the fate of a group of individuals all born in the same year. Of this group, or cohort, only a certain number of individuals will reach each age, and there is an age above which no individuals ever survive. Plotting the number of those members of the group that are still alive at each age results in a survivorship curve for the population. Survivorship curves are usually displayed on a semilogarithmic rather than an arithmetic scale. There are three general types of survivorship curves. Species such as humans and large mammals, which have fewer numbers of offspring but invest much time and energy in caring for their young (*K*-selected species), usually have a Type I curve. This relatively flat curve reflects low juvenile mortality, with most individuals living to old age. A constant probability of dying at any age, represented by the Type II curve, is evident as a straight line

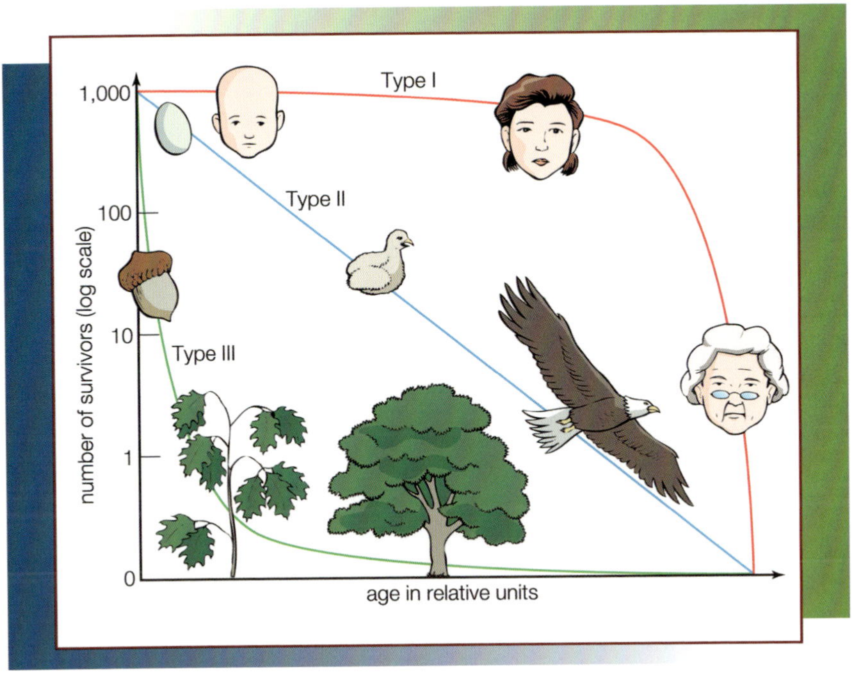

Survivorship curves. Encyclopædia Britannica, Inc.

decreasing over time toward zero. Certain lizards, perching birds, and rodents exhibit this type of survivorship curve. In some species that produce many offspring but provide little care for them (*r*-selected species), mortality is greatest among the youngest individuals. The Type III survivorship curve indicative of this life history is initially very steep but flattens out as those individuals who reach maturity survive for a relatively longer time; it is exhibited by animals such as many insects or shellfish. Many populations have survivorship patterns that are more complex than, or fall in between, these three idealized curves. For example, passerine birds (perching birds such as finches) commonly suffer high mortality during the first year of life and a lower, more constant rate of death in subsequent years.

Calculating Population Growth

Life tables also are used to study population growth. The average number of offspring left by a female at each age and the proportion of individuals surviving to each age can be used to evaluate the rate at which the size of the population changes over time. These rates are used by demographers and population ecologists to estimate human population growth and to evaluate the effect that conservation efforts have on endangered species.

The average number of offspring that a female leaves during her lifetime is called the net reproductive rate (Ro). If all females survived to the oldest possible age for that population, the net reproductive rate would simply be the sum of the average number of offspring left by females at each age. In real populations, however, some females die at every age. The net reproductive rate for a set cohort is obtained by multiplying the proportion of females surviving to each age (lx) by the average number of offspring produced at each age (mx) and then adding the products from all the age groups: $Ro = \Sigma lxmx$. A net reproductive rate of 1.0 indicates that a population is neither increasing nor decreasing but replacing its numbers exactly. Any number below 1.0 indicates a decrease in population, any number above indicates an increase. With a net reproductive rate of 2.101, the Galapagos cactus finch (*Geospiza scandens*) can double its size each generation.

The other value needed to calculate the rate at which the population can grow is the mean generation time (T). Generation time is the average interval between the birth of an individual and the birth of its offspring. To determine the mean generation time of a population, the age of the individuals (x) is multiplied by the proportion of females surviving to that age (lx) and the average number of offspring left by females at that age (*mx*). This calculation is performed for each age group, and the values are added together and divided by the net reproductive rate (*Ro*) to yield the result:

$$T = \frac{\Sigma x l_x m_x}{R_o}$$

The mean generation time of the Galapagos cactus finch is 6.08 years.

The value that is used by population biologists to calculate the rate of increase of populations is the intrinsic rate of natural increase (*r*), or the Malthusian parameter. Very simply, this rate can be understood as number of births minus number of deaths per generation time—in other words, the reproduction rate less the death rate. To derive this value using a life table, the natural logarithm of the net reproductive rate is divided by the mean generation time:

$$r = \frac{\ln R_o}{T}$$

Values above zero indicate that the population is increasing. The higher the value, the faster the growth rate. The Malthusian parameter can be used to compare growth rates of populations of a species that have different generation times. Some human populations have higher intrinsic rates of natural increase partially because individuals in those groups begin reproducing earlier than those in other groups. Mice have higher intrinsic rates of natural increase than elephants because they reproduce at a much earlier age and have a much shorter mean generation time.

If a population has an intrinsic rate of natural increase of 0, then it is said to have stable age distribution and is neither growing nor declining in numbers. A growing population has more individuals in the lower age classes than does a stable population, and a declining population has more individuals in the older age classes than does a stable population. Many human populations are currently undergoing population increase, far exceeding a stable age distribution. Although the human population has increased almost continuously throughout history, it has skyrocketed since the Industrial Revolution, primarily because of a drop in death rates. No other population has shown such steady growth.

The Regulation of Populations

Populations are not static. They rise and fall according to the availability of food resources, space, and competition. Ultimately, rising populations reach the numerical ceiling, or carrying capacity, of the environment they live in. Several individual populations may be connected to one another by dispersing individuals.

The Limits to Population Growth

There are two types of population growth. Exponential population growth occurs in situations where resources are effectively unlimited, whereas logistic growth manifests itself as the population approaches the environment's carrying capacity.

Exponential Population Growth

In an ideal environment, one that has no limiting factors, populations grow at a geometric rate or exponential rate. Human populations, in which individuals live and reproduce for many years and in which reproduction is distributed throughout the year, grow exponentially.

Exponential population growth can be determined by dividing the change in population size (QN) by the time interval (Qt) for a certain population size (N):

$$\frac{\Delta N}{\Delta t} = rN.$$

The growth curve of these populations is smooth and becomes increasingly steeper over time. The steepness of the curve depends on the intrinsic rate of natural increase for the population. Human population growth has been exponential during the 20th century. Much concern exists about the impact this growth will have, not only on the environment, but on humans as well. The World Bank projection for human population growth predicts that the human population will grow from 4.5 billion in 1995 to 10.2 billion in 2100. This estimate could be offset by thrcc population-control measures: (1) lower the rate of unwanted births, (2) lower the desired family size, and (3) raise the average age at which women begin to bear children or reduce the number of births below the level that would replace current human populations (e.g., one child per woman).

Insects and plants that live for a single year and reproduce once before dying are examples of organisms whose growth is geometric. In these species a population grows as a series of increasingly steep steps rather than as a smooth curve.

Logistic Population Growth

The geometric or exponential growth of all populations is eventually curtailed by limitation of food, competition for other resources, disease, or some other ecological factor. If growth is limited by resources such as food, the exponential growth of the population begins to slow as competition for those resources increases. The growth of the population eventually slows nearly to zero as the population reaches the carrying capacity (K) for the environment. The result is an S-shaped curve of population growth known as the logistic curve. It is determined by the equation:

$$\frac{\Delta N}{\Delta t} = rN\left(\frac{K - N}{K}\right).$$

HABITATS AND TERRITORIES

Habitats and territories are critical parts of an organism's existence. These terms, however, are sometimes confused with one another. A habitat is a place where an organism or a community of organisms lives, including all living and nonliving factors or conditions of the surrounding environment. A host organism inhabited by parasites is as much a habitat as a terrestrial place such as a grove of trees or an aquatic locality such as a small pond. Microhabitat is a term for the conditions and organisms in the immediate vicinity of a plant or animal.

In contrast, a territory is any area defended by an organism or a group of similar organisms for such purposes as mating, nesting, roosting, or feeding. Most vertebrates and some invertebrates, such as arthropods, including insects, exhibit territorial behaviour. Possession of a territory involves aggressive behaviour and thus contrasts with the home range, which is the area in which the animal normally lives. Home range is not associated with aggressive behaviour, although parts of the home range may be defended: in this case the defended part is the territory. The type of territory varies with the social behaviour and environmental and resource requirements of the particular species and often serves more than one function, but whatever the type, the territory acts as a spacing mechanism and a means of allocating resources among a segment of the population and denying it to others. Some authorities also consider plants or animals that secrete repulsive chemicals into their immediate environments to be territorial, because the substances space individuals of the species apart from one another.

Population Fluxes

As stated earlier, populations rarely grow smoothly up to the carrying capacity and then remain there. Instead, fluctuations in population numbers are the norm. In a few species, such as snowshoe hares, lemmings, lynx, and Arctic foxes, populations show regular cycles of increase and decrease spanning a number of years. The causes of these fluctuations are still under debate by population ecologists, and no single cause may provide an explanation for every species. Most major hypotheses link regular fluctuations in population size to factors that are dependent on the density of the population, such as the availability of food.

Population ecologists commonly divide the factors that regulate the size of populations into density-dependent and density-independent factors. Density-independent factors, such as weather and climate, affect the same proportion of individuals in a population regardless of population density. In contrast, the effects of density-dependent factors intensify as the population increases in size. For example, some diseases spread faster in dense populations than in sparse populations. Similarly, competition for food and other resources rises with density and affects an increasing proportion of the population. The dynamics of most populations are influenced by both density-dependent and density-independent factors, and the relative effects of the factors vary among populations.

Because many factors influence population size, erratic variations in number are more common than regular cycles of fluctuation. Some populations undergo unpredictable irruptions in numbers, sometimes temporarily increasing by 10 or 100 times over a few years, only to follow with a similarly rapid crash. Migratory grasshoppers in the arid

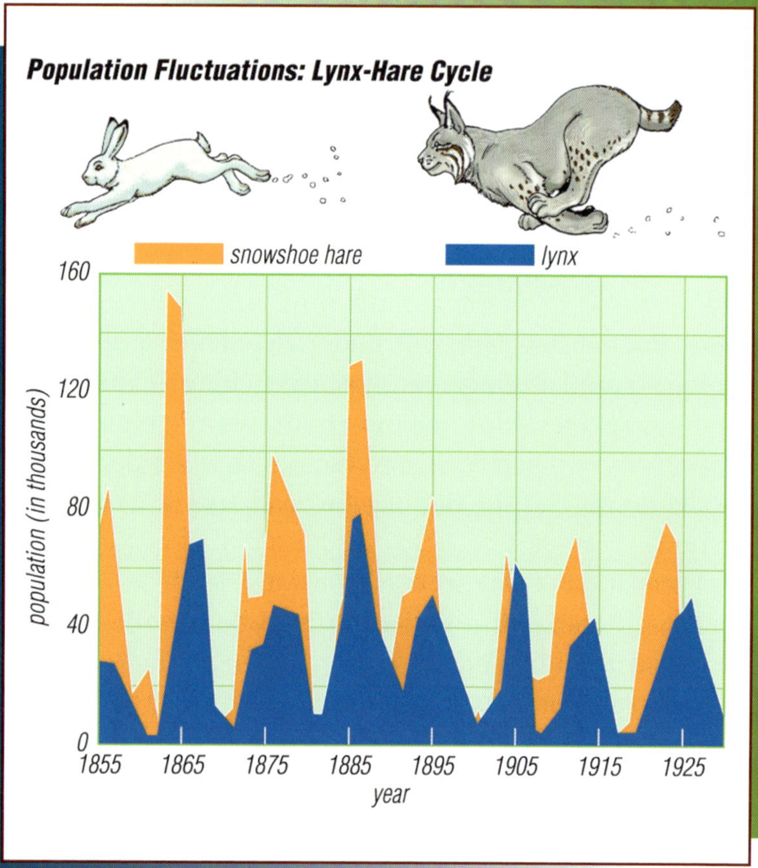

Cyclical fluctuations in the population density of the snowshoe hare and its effect on the population of its predator, the lynx. The graph is based on data derived from the records of the Hudson's Bay Company. Encyclopædia Britannica, Inc.

parts of Africa multiply to such a level that their numbers can blacken the sky overhead; similar upsurges occurred in North America before the 20th century. Some forest insects, such as the gypsy moths that were introduced to North America, are highly irruptive. As with species that fluctuate more regularly, the causes behind population irruptions are not fully known and are unlikely to have a single explanation that applies to all species.

The size of other populations varies within tighter limits. Some fluctuate close to their carrying capacity, others fluctuate below this level, held in check by various ecological factors, including predators and parasites. The tremendous expansion of many populations of weeds and pests that have been released into new environments in which their enemies are absent suggests that predators, grazers, and parasites all contribute to maintaining the small sizes of many populations. To control the explosive proliferation of these species, biological control programs have been instituted. With varying degrees of success, parasites or pathogens inimical to the foreign species have been introduced into the environment. The European rabbit (*Oryctolagus cuniculus*) was introduced into Australia in the 1800s, and its population grew unchecked, wreaking havoc on agricultural and pasture lands. The myxoma virus subsequently was released among the rabbit populations and greatly reduced them. Populations of the prickly pear cactus (*Opuntia*) in Australia and Africa grew unbounded until the moth borer (*Cactoblastis cactorum*) was introduced. Many other similar attempts at biological control have failed, illustrating the difficulty in pinpointing the factors involved in population regulation.

Metapopulations

Although the dynamics and evolution of a single closed population are governed by its life history, populations of many species are not completely isolated and are connected by the movement of individuals (immigration and emigration). Consequently, the dynamics and evolution of many populations are determined by both the population's life history and the patterns of movement of individuals between populations. Regional groups of interconnected populations are called metapopulations. These metapopulations are, in turn, connected to one another over broader geographic ranges. The distribution of the metapopulation structure of the perennial herb *Clematis fremontii* variety *Riehlii* in Missouri spans 1,129 square kilometres (436 square miles). There is, therefore, a hierarchy of population structure from local populations to metapopulations to broader geographic groups of populations and eventually up to the worldwide collection of populations that constitute a species.

As local populations within a metapopulation fluctuate in size, they become vulnerable to extinction during periods when their numbers are low. Extinction of local populations is common in some species, and the regional persistence of such species is dependent on the existence of a metapopulation. Hence, elimination of much of the metapopulation structure of some species can increase the chance of regional extinction of species.

The structure of metapopulations varies among species. In some species one population may be particularly stable over time and act as the source of recruits into other, less stable populations. For example, populations of the checkerspot butterfly (*Euphydryas editha*) in California have a metapopulation structure consisting of a number of small satellite populations that surround a large source population on which they rely for new recruits. The satellite populations are too small and fluctuate too much to maintain themselves indefinitely. Elimination of the source population from this metapopulation would probably result in the eventual extinction of the smaller satellite populations.

In other species, metapopulations may have a shifting source. Any one local population may temporarily be the stable source population that provides recruits to the more unstable surrounding populations. As conditions change, the source population may become unstable, as when disease increases locally or the physical environment deteriorates. Meanwhile, conditions in another population that had previously been unstable might improve, allowing this population to provide recruits.

Overall, the population ecology and dynamics of all species is a complex result of their genetic structure, the life histories of the individuals, fluctuations in the carrying capacity of the environment, the relative influences of all the different kinds of density-dependent and density-independent factors that limit population growth, the spatial distribution of individuals, and the pattern of movement between populations that determines metapopulation structure. It is therefore not surprising that there are often great fluctuations in the numbers of individuals in local populations and that the long-term persistence of species may often require the conservation of many, rather than a few, populations.

CHAPTER 5

COMMUNITY ECOLOGY

Community ecology, or synecology, is the study of the organization and functioning of communities, which are assemblages of interacting populations of the species living within a particular area or habitat. As populations of species interact with one another, they form biological communities. The number of interacting species in these communities and the complexity of their relationships exemplify what is meant by the term "biodiversity." Structures arise within communities as species interact, and food chains, food webs, guilds, and other interactive webs are created. These relationships change over evolutionary time as species reciprocally adapt to one another through the process of coevolution. The overall structure of biological communities, the organization of interspecific interactions, and the effects the coevolutionary process has on the biological community are described in this chapter.

THE BIOTIC ELEMENTS OF COMMUNITIES

Although community structure depends largely on the physical environment, there are four important biotic phenomena that hold communities together. Biotic communities are characterized by the number of trophic levels they can support and how energy moves from one to another. The community is also guided by the establishment of food chains. Often perched at the top of food chains, some members of the community called keystone species have a disproportionate influence over how nutrients and energy move through the ecosystem. Lastly, the habits of groups of species may converge to exploit an ecosystem in similar ways. Such species often compete or cooperate with one another, greatly expanding the number of ways in which species interact and thus contributing to the community's stability.

Trophic Pyramids and the Flow of Energy

All biological communities have a basic structure of interaction that forms a trophic pyramid. The trophic pyramid is made up of trophic levels, and food energy is passed from one level to the next along the food chain. The base of the pyramid is composed of species called autotrophs, the primary producers of the ecosystem. They do not obtain energy and nutrients by eating other organisms. Instead, they harness solar energy by photosynthesis (photoautotrophs) or, more rarely, chemical energy by oxidation (chemoautotrophs) to make organic substances from inorganic ones. All other organisms in the ecosystem are consumers called heterotrophs, which either directly or indirectly depend on the producers for food energy.

Within all biological communities, energy at each trophic level is lost in the form of heat (as much as 80 to 90 percent), as organisms expend energy for metabolic processes such as staying warm and digesting food. The higher the organism is on the trophic pyramid, the less energy is available to it. Herbivores and detritivores (primary consumers) have less available energy than plants, and the carnivores that feed on herbivores and detritivores (secondary consumers) and those that eat other carnivores (tertiary consumers) have the least amount of available energy.

The organisms that make up the base level of the pyramid vary from community to community. In terrestrial communities, multicellular plants generally form the base of the pyramid, whereas in freshwater lakes a combination of multicellular plants and single-celled algae constitute the first trophic level. The trophic structure of the ocean is built on the plankton known as krill. There are some exceptions to this general plan. Many freshwater streams have detritus rather than living plants as their energy base. Detritus is composed of leaves and other plant parts that fall into the water from surrounding terrestrial communities. It is broken down by microorganisms, and the microorganism-rich detritus is eaten by aquatic invertebrates, which are in turn eaten by vertebrates.

The most unusual biological communities of all are those surrounding hydrothermal vents on the ocean floor. These vents result from volcanic activity and the movement of continental plates that create cracks in the seafloor. Water seeps into the cracks, is heated by magma within the Earth's mantle, becomes laden with hydrogen sulfide, and then rises back to the ocean floor. Sulfur-oxidizing bacteria (chemoautotrophs) thrive in the warm, sulfur-rich water surrounding these cracks. The bacteria use reduced sulfur as an energy source for the fixation of carbon dioxide. Unlike all other known biological communities on Earth, the energy that forms the base of these deep-sea communities comes from chemosynthesis rather than from photosynthesis. The ecosystem is thus supported by geothermal rather than solar energy.

Some species surrounding these vents feed on these bacteria, but other species have formed long-term, reciprocally beneficial relationships (mutualistic symbioses) with sulfur bacteria. These species harbour the chemoautotrophic bacteria within their bodies and derive nutrition directly from them. The biological communities surrounding these vents are so different from those in the rest of the ocean that since the 1980s, when biological research of these vents began, about 200 new species have been described, and there are many more that remain undescribed (i.e., not formally described and given scientific names). Among the described species there are at least 75 new genera, 15 new families, one new order, one new class, and even one new phylum.

Food Chains and Food Webs

Because all species are specialized in their diets, each trophic pyramid is made up of a series of interconnected feeding relationships called food chains. Most food chains consist of three or four trophic levels. A typical sequence may be plant, herbivore, carnivore, top carnivore; another sequence is plant, herbivore, parasite of the herbivore, and parasite of the parasite. Many herbivores, detritivores, carnivores, and parasites, however, eat more than one species, and a large number of animal species eat different foods at different stages of their life histories. In addition, many species eat both plants and animals and therefore feed at more than one trophic level. Consequently, food chains combine into highly complex food webs. Even a simplified food web can show a complicated network of trophic relationships.

Keystone Species

Even a fully constructed food web, however, can provide only a superficial and static view of the structure of biological communities. Not all the relationships between species are of equal importance in the dynamics and evolution of populations and the organization of communities. Food webs include both strong and weak interactions between species, and these differences in interaction

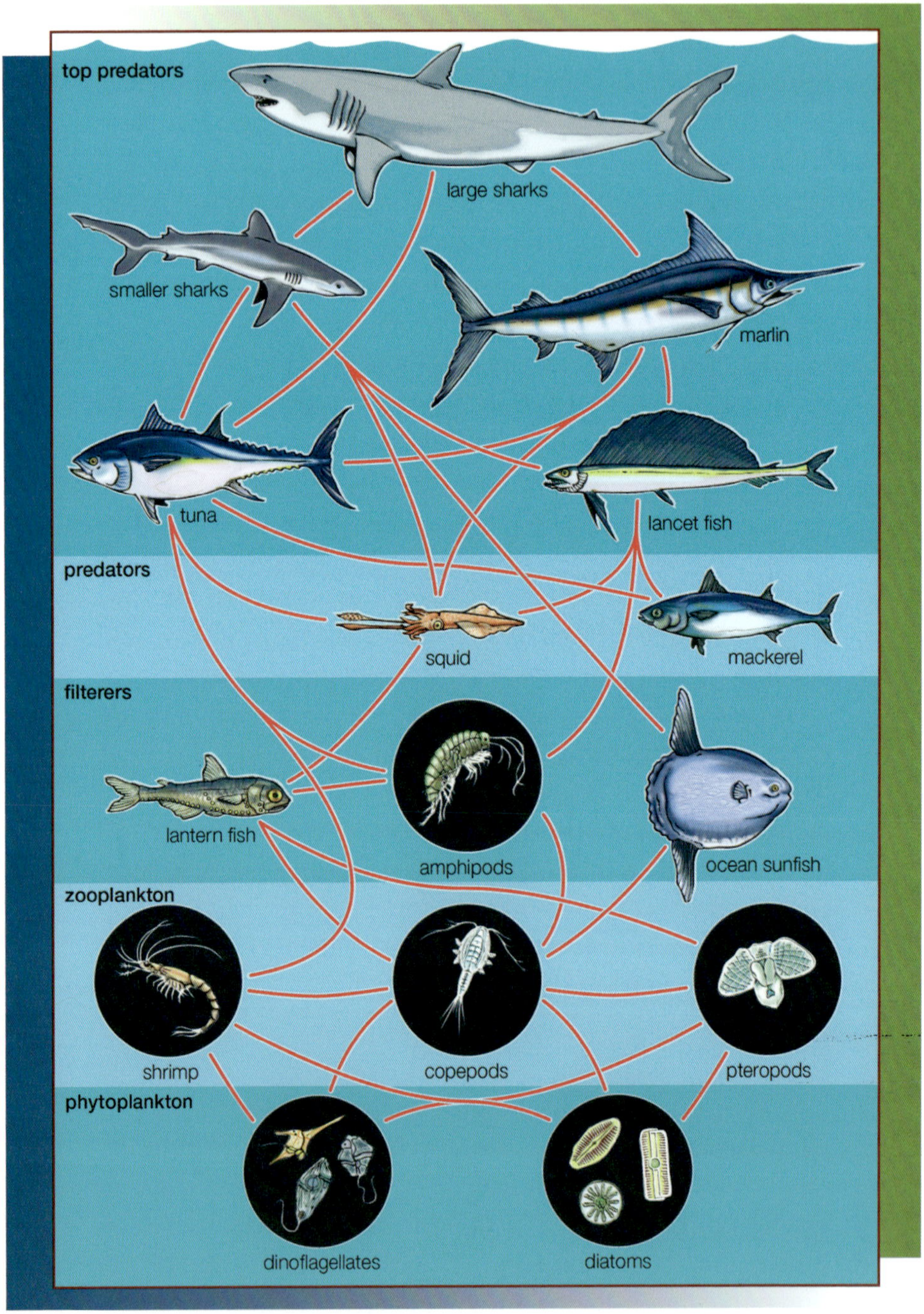

Generalized aquatic food web. Parasites, among the most diverse species in the food web, are not shown. Encyclopædia Britannica, Inc.

strength influence the organization of communities. Some species, called keystone species, have a disproportionately large effect on the communities in which they occur. They help to maintain local diversity within a community either by controlling populations of species that would otherwise dominate the community or by providing critical resources for a wide range of species.

The starfish *Pisaster ochraceus* is a keystone species in the rocky marine intertidal communities off the northwest coast of North America. This predatory starfish feeds on the mussel *Mytilus californianus* and is responsible for maintaining much of the local diversity of species within certain communities. When the starfish have been removed experimentally, the mussel populations have expanded rapidly and covered the rocky intertidal shores so exclusively that other species cannot establish themselves. Consequently, the interaction between *Pisaster* and *Mytilus* supports the structure and species diversity of these communities. In other communities in which *Pisaster* occurs, however, the starfish has little overall effect on the structure of the community. Therefore, a species can be a keystone species in some communities but not in others.

In some forest communities in tropical America, figs and a few other plants act as keystone species but in a very different manner from the starfish *Pisaster*. Figs bear fruit year-round in some of these forest communities, and a large number of birds and mammals rely heavily on this small group of plant species during the times of the year when other food resources are scarce. Without figs, many species would disappear from the community.

Guilds and Interaction Webs

Most communities contain groups of species known as guilds, which exploit the same kinds of resources in comparable ways. The name "guild" emphasizes the fact that these groups are like associations of craftsmen who employ similar techniques in plying their trade. Guilds may consist of different insect species that collect nectar in similar ways, various bird species that employ corresponding insect-foraging techniques, or diverse plant species that have evolved comparable floral shapes with which they attract the same group of pollinators.

Guilds often are composed of groups of closely related species that all arose from a common ancestor. They exploit resources in similar ways as a result of their shared ancestry. Hence, several species within a single genus may constitute a guild within a community. A less common but not unknown occurrence is for unrelated species to make up a guild.

Because members of a guild engage in similar activities, it is not surprising that they are often competitors for the resources they share, especially when those resources are scarce. This competition among guilds emphasizes the fact that, in addition to food webs, the structure of the community is built on other types of interaction. Species not only eat one another, they compete for resources, forging a variety of interspecific interactions. Many species also interact cooperatively to search for food or avoid predators. These and other nontrophic relationships between species are as important as food chains and food webs in shaping the organization of biological communities.

PATTERNS OF COMMUNITY STRUCTURE

Ecological communities that persist often contain a diverse assemblage of species. Although the change that comes to ecological communities is often a gradual process, as one season gives way to another and young trees grow tall over many years, storms, fire, disease, or other

NICHES AND COMPETITIVE EXCLUSION

A niche is a compilation of all of the interactions of a species with the other members of its community, including competition, predation, parasitism, and mutualism. A variety of abiotic factors, such as soil type and climate, also define a species' niche. Each of the various species that constitute a community occupies its own ecological niche. Informally, a niche is considered the "job" or "role" that a species performs within nature.

The principle of competitive exclusion—which is also called Gause's principle, or Grinnell's axiom (after G.F. Gause, a Soviet biologist, and J. Grinnell, an American naturalist, who first clearly established it)—states that in competition between species that seek the same ecological niche, one species survives while the other expires under a given set of environmental conditions. The result is that each species occupies a distinct niche.

disturbances occur periodically. Occasionally, these forces are so severe that they test the resilience of the community. However, if such communities are biologically diverse, they often recover quickly from disturbances.

Ecological Succession

The structure of communities is constantly changing. All communities are subject to periodic disturbances, ranging from events that have only localized effects, such as the loss of a tree that creates a gap in the canopy of a forest, to those that have catastrophic consequences, which include wildfires that sweep across vast landscapes or storms that pound immense stretches of shoreline. Each new disturbance within a landscape creates an opportunity for a new species to colonize that region. New species also alter the character of the community, creating an environment that is suitable to even newer species. By this process, known as ecological succession, the structure of the community evolves over time.

Two different types of succession, primary and secondary, have been distinguished. Primary succession occurs in essentially lifeless areas—regions in which the soil is incapable of sustaining life as a result of such factors as lava flows, newly formed sand dunes, or rocks left from a retreating glacier. Secondary succession occurs in areas where a community that previously existed has been removed. It is typified by smaller-scale disturbances that do not eliminate all life and nutrients from the environment. Events such as a fire that sweeps across a grassland or a storm that uproots trees within a forest create patches of habitat that are colonized by early successional species. Depending on the extent of the disturbance, some species may survive, other species may be recolonized from nearby habitats, and others may actually be released from a dormant condition by the disturbance. For example, many plant species in fire-prone environments have seeds that remain dormant within the soil until the heat of a fire stimulates them to germinate.

Primary and secondary succession both create a continually changing mix of species within communities as disturbances of different intensities, sizes, and frequencies alter the landscape. The sequential progression of species during succession, however, is not random. At every stage certain species have evolved life histories to exploit the particular conditions of the community. This situation imposes a partially predictable sequence of change in the species composition of communities during succession. Initially, only a small number of species from surrounding habitats are capable of thriving in a disturbed habitat. As new plant species take hold, they modify the habitat by altering such things as the amount of shade on the ground or the mineral composition of the soil. These changes allow other species that are better suited to this modified habitat to succeed the old species. These newer species are superseded, in turn, by still newer species. A similar succession of animal species occurs, and interactions between plants, animals, and environment influence the pattern and rate of successional change.

Community Stratification and Gradation

Community structure can become stratified both vertically and horizontally during the process of succession as species become adapted to their habitat. Gradations in environmental factors such as light, temperature, or water are responsible for this fractionation. The vertical stratification that occurs within forests results from the varying degrees of light that the different strata receive: the taller the plant and the more foliage it produces, the more light it can intercept. Three or more vertical strata of plants—an herb layer, a shrub layer, a small tree layer, and a canopy tree layer—often are found in a forest. Animals are affected by this stratification of plant life. Although they can move from one layer to another quite easily, they often adhere closely to a specific layer for foraging, breeding, or other activities.

Horizontal patterns among species also can emerge from gradients in the physical environment. Differences in the amount of water or nutrients over a region can affect the distribution of animal and plant species. On a mountain, plant and animal species vary at different elevations as well as among the north, south, east, and west slopes. Drastic differences in certain factors over a very short distance can create sharp boundaries between communities, whereas gradual differences can produce a more integrated flow of species. These gradients help to maintain regional biodiversity.

Ecotones

Ecosystems are almost always a patchwork of communities that exist at different successional stages. The sizes, frequencies, and intensities of disturbances differ among ecosystems, creating differences in what is called the patch dynamics of communities. Along the edges of each of the patches are areas called ecotones. These junction zones often contain species of each of the overlapping communities as well as some species that have become adapted specifically for living in these zones. In many cases, the number of species and the population density are greater within the ecotone than in the surrounding communities, a phenomenon known as the edge effect.

In North America the parasitism of bird nests by brown-headed cowbirds (*Molothrus ater*) is particularly frequent in ecotones between mature forests and earlier successional patches. Cowbirds lay their eggs in the nests of other birds and are active mainly in early successional patches. Forest birds whose nests are deep within the interior of mature forests are less likely to be attacked than those within ecotones. The cutting of mature forests has increased the extent of ecotones, concomitantly increasing the rate of cowbird parasitism across North America.

Ecological Niches

An ecological niche encompasses the habits of a species. Essentially, it refers to the way a species relates to, or fits in with, its environment. As a species adapts to the physical parameters and biota within the community, natural selection favours the development of specialized features that allow the species to uniquely exploit the surrounding resources. Physical conditions of the region—such as temperature, terrain, or nutrient availability—help to mold the niche, and biological constraints such as predation, competition, or lack of resources limit the ways in which a species exploits its environment. For example, plant species differ in their requirements for light, nutrients, and microorganisms, as well as in their ability to fend off competitors and herbivores. Herbivore species can eat only a subset of the plants available within a community, and predators can capture only some of the many potential prey species. Thus the species "carves out" a niche for itself in the community.

An example of one such niche is that of the endangered Kirtland's warbler (*Dendroica kirtlandii*) found in North America. It nests only among young jack pines (*Pinus banksiana*) that are 2 to 4 metres (6.5 to 13 feet) tall and grow in homogenous stands. These trees are exposed to periodic fires, necessary for germination of the jack pine seeds. These fires also continuously provide extensive new regions of young trees, allowing the warblers to shift their nesting sites over the years to remain within stands of jack pine that are of the preferred height.

The niche of a species evolves as physical and biological factors in the community change—provided that such changes are slow enough to allow species to adapt to them. The main constraint on this evolution is that no two species in a community can have the same niche. Specialized modes of existence thus provide a selective advantage to coexistent species, offsetting direct competition for available resources.

Biodiversity and the Stability of Communities

As species adapt to one another and to their communities, they form niches and guilds. The development of more complex structures allows a greater number of species to coexist with one another. The increase in species richness and complexity acts to buffer the community from environmental stresses and disasters, rendering it more stable.

Community Equilibrium and Species Diversity

In some environments, succession reaches a climax, producing a stable community dominated by a small number of prominent species. This state of equilibrium, called the climax community, is thought to result when the web of biotic interactions becomes so intricate that no other species can be admitted. In other environments, continual small-scale disturbances produce communities that are a diverse mix of species, and any species may become dominant. This nonequilibrial dynamic highlights the effects that unpredictable disturbances can have in the development of community structure and composition. Some species-rich tropical forests contain hundreds of tree species within a square kilometre. When a tree dies and falls to the ground, the resultant space is up for grabs. Similarly, some coral reefs harbour hundreds of fish species, and whichever species colonizes a new disturbance patch will be the victor. With each small disturbance, the bid for supremacy begins anew.

Diverse communities are healthy communities. Long-term ecological studies have shown that species-rich communities are able to recover faster from disturbances than species-poor communities. Species-rich grasslands in the Midwestern United States maintain higher primary productivity than species-poor grasslands. Each additional species lost from these grasslands has a progressively greater effect on the drought-resistance of the community. Similarly, more diverse plant communities in Yellowstone National Park show greater stability in species composition during severe drought than less diverse communities. And, in the Serengeti grassland of Africa, the more diverse communities show greater stability of biomass through the seasons and greater ability to recover after grazing.

The relationship between species diversity and community stability highlights the need to maintain the greatest richness possible within biological communities. A field of weeds containing species only recently introduced to the community is quite different from a rich interactive web of indigenous species that have had the time to adapt to one another. Undisturbed species-rich communities have the resilience to sustain a functioning ecosystem upon which life depends. These communities also are better able to absorb the effects of foreign species, which may be innocently introduced but which can wreak much ecological and economic havoc in less stable communities. The tight web of interactions that make up natural biological communities sustains both biodiversity and community stability.

The Biogeographic Aspects of Diversity

Biogeography is the study of species distribution in an area. Because islands provide a controlled area for study, they have been used to observe the factors that affect species diversity. Three variables that determine the rate of colonization of an island are the size of the island, the distance between the island and other islands or the mainland, and the number of species inhabiting the surrounding lands. The theory of island biogeography is based on this information, which can help predict the number of species that will occur on a given island. It also can be used to explain the species diversity of "islands" on land, such as mountaintops, lakes, and forest fragments left after an area has been logged. Where immigration and extinction rates are equal, the theory of island biogeography states that the number of species is proportional to the size of the island and inversely proportional to the distance of the island from the mainland.

Six months after the eruption of a volcano on the island of Surtsey off the coast of Iceland in 1963, the island had been colonized by a few bacteria, molds, insects, and birds. Within about a year of the eruption of a volcano on the island of Krakatoa in the tropical Pacific in 1883, a few grass species, insects, and vertebrates had taken hold. On both Surtsey and Krakatoa, only a few decades had elapsed before hundreds of species reached the islands. Not all species are able to take hold and become permanently established, but eventually the island communities stabilize into a dynamic equilibrium.

THE INTERACTIONS BETWEEN SPECIES AND COMMUNITY ORGANIZATION

The interactive relationships that arise between populations of different species form the interactive web of communities. These interactions range from antagonistic to cooperative and have either positive, negative, or neutral effects on the species involved. In antagonistic relationships the interaction is detrimental to individuals of either one or both species. In commensal relationships (commensalism) one species benefits while the other remains unaffected, whereas in mutualistic relationships (mutualism) both species benefit. The organization and stability of biological communities results from the mix of these different kinds of interaction.

These relationships between species are not static. They evolve as natural selection continually shapes and reshapes them. The defenses and counterdefenses seen in the relationships between hosts and parasites, or between prey and predators, are snapshots of one point in time during the ongoing process of the evolution of interactions. As interactions between species evolve, relationships may

SYMBIOSIS

A symbiosis describes any of several living arrangements between members of two different species, including mutualism, commensalism, and parasitism. Both positive (beneficial) and negative (unfavourable to harmful) associations are therefore included, and the members are called symbionts. Any association between two species populations that live together is symbiotic, whether the species benefit, harm, or have no effect on one another.

Clownfish (Amphiprion) *amid sea anemones* (Heteractis magnifica), *Great Barrier Reef, Australia. The two species protect each other from predators in mutual symbiosis.* Jan Derk

shift from antagonism to commensalism to mutualism. As a result, the organization of biological communities is no more fixed than are the characteristics of the species or their environments. Charles Darwin called this ever-changing mix of species and their interactions the "entangled bank" and stressed its importance in the evolutionary process.

The Role of Mutualism

In attempting to unravel Darwin's entangled bank and understand how these interactions form the basic structure of communities, many popular accounts of community ecology focus on extravagant antagonistic displays between species. Although aggressive behaviours are important interspecific interactions, the amount of attention that is focused on them may create the incorrect impression that they are more important than other types of interaction. Mutualistic interactions between species are just as integral to the organization of biological communities as antagonistic relationships, with some mutualistic interactions forming the most basic elements of many communities.

In many mutualistic relationships, one species acts as the host, and the other plays the role of visitor or resident. Plants are hosts for insects that pollinate them or eat their fruit and for microorganisms that attach themselves to their roots. In other mutualisms, such as flocks of birds that include a mixture of species, no species acts as host. Mutualisms also vary in the benefits the participants derive from the interaction. An individual may gain food, protection from enemies, a nesting site, or a combination of benefits. These benefits may vary from one population to another, thereby causing mutualistic relationships that exist between the same species to evolve in different directions in different populations.

The Pervasiveness of Mutualism

Some mutualistic relationships are so pervasive that they affect almost all life-forms. The root systems of most terrestrial plant species form complex associations with the soil microorganisms. These mycorrhizal associations aid the plant in taking up nutrients. In some environments,

MUTUALISM

A mutualism is an association between organisms of two different species in which each is benefited. Mutualistic arrangements are most likely to develop between organisms with widely differing living requirements. The partnership between nitrogen-fixing bacteria and leguminous plants is an example, as is the association between cows and rumen bacteria (the bacteria live in the digestive tract and help digest the plants eaten by the cow). The associations between tree roots and certain fungi are often mutualistic.

Intestinal flagellated protozoans and termites exhibit obligative mutualism, a strict interdependency, in which the protozoans digest the wood ingested by the termites. Neither partner can survive under natural conditions without the other.

Acacia ants (*Pseudomyrmex ferruginea*) inhabit the bull-horn acacia (*Acacia cornigera*), upon which they obtain food and shelter. The acacia depends on the ants for protection from browsing animals, which the ants drive away. Neither member can survive successfully without the other, also exemplifying obligative mutualism.

The yucca moth is dependent on the yucca plant and vice versa: the moth acts as pollinator at the same time that she lays her eggs in the seed pods of the yucca. The larvae hatch and feed on some but not all the seeds. Both organisms benefit: the plant is pollinated, and the moth has a source of food for its larvae.

many plants cannot become established without the aid of associated mycorrhizae. In another relationship, legumes rely on nodule-forming associations between their roots and microorganisms to fix nitrogen, and these nitrogen-fixing plants are in turn crucial to the process of succession in biological communities.

Mutualistic associations between animals and microorganisms are equally important to the structure of communities. Most animals rely on the microorganisms in their gut to properly digest and metabolize food. Termites require cellulose-digesting microorganisms in their gut to obtain all possible nourishment that their diet of wood can provide.

At an even more fundamental level, the very origin of eukaryotic cells (those cells having a well-defined nucleus and of which higher plants and animals, protozoa, fungi, and most algae consist) appears to have resulted from an association with various single-celled species: the mitochondria and chloroplasts that occur in eukaryotic cells are thought to have originated as separate organisms that took up residence inside other cells. Eventually neither organism was able to survive without the other—a situation called obligative symbiosis.

In many terrestrial environments, mutualisms between animals and plants are central to the organization of biological communities. In some tropical communities, animals pollinate the flowers and disperse the seeds of almost every woody plant. In turn, a large proportion of animals rely on flowers or fruits for at least part of their diet. Leaf-cutting ants, an important species in neotropical forest communities, prepare cut leaves as a substrate on which to grow the specialized fungus gardens on which they feed. Thousands of plant species produce extrafloral nectaries on their leaves or petioles to attract many kinds of ants, which feed on the nectar and kill insect herbivores that they encounter on the plants.

The Evolution of Mutualism

Although mutualisms benefit all species involved in a relationship, they are built on the same genetically selfish principles as antagonistic interactions. In fact, many mutualisms appear to have evolved from antagonistic interactions. No species behaves altruistically to promote the good of another species. Mutualisms evolve as species that come in contact manipulate each other for their own benefit. Plants evolve particular mixtures and concentrations of nectar to tempt pollinators to behave in ways that maximize pollination. Purely for their own advantage, pollinators visit plants and navigate among them to harvest nectar or pollen in the most efficient way possible. Their concern is not with how well they function as pollinators for the plants but rather with what they can extract from the plants. Mutualism results whenever the selfish activities of species happen to benefit each of them. Natural selection continues to reshape these relationships as each species evolves its ability to exploit the other.

Mutualism and Cheaters

Because mutualisms develop through the manipulation of other species, they are always susceptible to invasion by "cheaters," those organisms that can exploit an existing relationship without reciprocating an advantage. Theft of a resource is one type of crime in which a cheater engages. Some plants, for example, have coevolved with particular pollinators. The flowers of these plants have deep corollas (inner sets of leaves of the flower constituting an inner chamber) with nectar at the bottom that is accessible only to their pollinators that have long tongues or bills specialized for this purpose. Some short-tongued bees and short-billed hummingbirds, however, have developed their own adaptations—they extract nectar by piercing the base of these long corollas. Another form of cheating involves mimicking the appearance of one species to subvert an existing mutualistic association. This subversion has occurred between cleaner fish and their hosts. Cleaner fish are highly specialized fish that pick parasites off the skin of other fish. Host fish arrive at specific sites where they present themselves to the cleaner fish that groom them. Other fish have evolved to resemble the cleaner fish, but, rather than search for parasites, these imposters take a bite out of the host fish.

Other cheaters use different deceptive devices to exploit mutualistic interactions. Crab spiders, phymatid bugs, and some praying mantids use flowers as places to wait for prey: they have evolved a camouflage that allows them to hide from the pollinating insects they feed on. Rather than construct a web or search through the vegetation to capture prey, this type of predator merely remains frozen on a flower until an unsuspecting pollinator stumbles into its clutches.

As cheaters evolve to exploit a mutualism, they can cause the symbiotic relationship itself to break down unless new ways of thwarting the pretenders evolve within the host population.

Community Structure and the Spread of Mutualism

As mutualisms spread within biological communities over evolutionary time, they make possible new lifestyles that rely on the availability of a number of mutualistic species. Once fleshy fruits had evolved in many plant species and had begun to occur together within communities, bird species evolved that were specialized physiologically to feed on fruits year-round rather than as a short-term seasonal addition to their diet. Resplendent quetzals (*Pharomachrus mocinno*) and oilbirds (*Steatornis caripensis*) have evolved in tropical American forests that have a succession of fruit species throughout the year. These highly specialized birds feed almost exclusively on fruits, supplying fruit even to their nestlings, and hence are called frugivores. To maintain this year-round diet of fruit, resplendent quetzals consume at least 43 fruit species from 17 plant families, and oilbirds eat at least 36 fruit species from 10 plant families. Similarly, hummingbirds, social bees such as honeybees, and other species that feed on nectar (nectarivores) and have life spans longer than the flowering time of one plant species have mutualistic relationships with a succession of pollinating species in order to survive.

This reliance on a succession of species by some frugivores and nectarivores is one reason that the piecemeal extinction of one plant species from biological communities, so common in recent decades, has such potentially disastrous consequences. The mutualisms between nectarivores and flowers, and between frugivores and fruits, are not just extraneous additions to the organization of biological communities. They are central relationships, because they ensure that the next generation of plants is produced and distributed throughout the landscape. The local extinction of a seemingly obscure plant, however, could easily lead to the local extinction of frugivores and nectarivores if these animals rely on that plant during times of scarcity of alternative plants. The importance of the seasonal succession of flowering and fruiting plant species and their associated nectarivores and frugivores to maintaining the normal functioning of terrestrial communities is only beginning to be appreciated.

Antagonism

Although mutualisms are common in all biological communities, they occur side by side with a wide array of antagonistic interactions. As life has evolved, natural selection has favoured organisms that are able to efficiently extract energy and nutrients from their environment. Because organisms are concentrated packages of energy and nutrients in themselves, they can become the objects of antagonistic interactions. Moreover, because resources often are limited, natural selection also has favoured the ability of organisms to compete against one another for them. The result has been the evolution of a great diversity of lifestyles. This diversity can be categorized in any number of ways, but the edges of all the categories blend with one another. Evolution continues to mix all the different kinds of interspecific interactions into novel ways of life.

One way of understanding the diversity of antagonistic interactions is through the kinds of hosts or prey that species attack. Carnivores attack animals, herbivores attack plants, and fungivores attack fungi. Other species are omnivorous, attacking a wide range of plants, animals, and fungi. Regardless of the kinds of foods they eat, however, there are some general patterns in which species interact. Parasitism, grazing, and predation are the three major ways in which species feed on one another. The parasite lives on and feeds off its host, usually decreasing the host's ability to survive but not killing it outright. Grazing species are not as closely tied to their food source as parasites and often vary their diet between two or more species without directly killing them. Predators, however, capture and kill members of other species for food.

Parasitism

Parasitism is thought to be the most common way of life, and parasitic organisms may account for as many as half of all living species. Examples include pathogenic fungi and bacteria, plants that tap into the stems or roots of other plants, insects that as larvae feed on a single plant, and parasitic wasps. Parasites live in or on a single host throughout either a stage in their lives or their entire life span, thereby decreasing the survival or reproduction of their hosts. This lifestyle has arisen many times throughout evolution. The most species-rich groups of organisms are parasites, which, in becoming specialized to live off their hosts alone, eventually become genetically distinct from their species, sometimes to the degree that they are considered a new species (speciation).

One common type of parasite is the parasitoid, an insect whose larvae feed and develop within or on the bodies of other arthropods. Each parasitoid larva develops on a single individual and eventually kills that host. Most parasitoids are wasps, but some flies and a small number of beetles, moths, lacewings, and even one caddisfly species have evolved to be parasitoids. Parasitoids alone number about 68,000 named species, and most parasitoids have yet to be named and described. Realistic estimates of the total number of described and undescribed parasitoid species are about 800,000.

The number of species of insects that develop as nymphs or larvae on a single plant host may outnumber the parasitoids. There is currently a grcat deal of debate concerning the number of species worldwide, and this debate centres on the number of plant-feeding insect species, many of which inhabit the canopies of tropical trees. These species have been almost impossible to collect until recently when techniques allowing access to the canopies have been adapted from mountain-climbing methods. All ecologists and systematists working on these estimates agree that there are at least a few million plant-feeding insect species, but the estimates range from 2 to 30 million.

Estimates of the number of pathogenic fungi, parasitic nematodes, and other parasitic groups also have increased as ecological and molecular studies are revealing many previously unrecognized species. Continued work on biodiversity worldwide will allow better estimates to be made of the Earth's inventory of species, which is a major prerequisite for understanding the role of parasites in the organization of communities and in the conservation of diversity.

It is now evident that the parasitic lifestyle often favours extreme specialization to a single host or a small group of hosts. Living for a long period of time on a single host, a parasite must remain attached within or on its host, avoid the defenses of its host, and obtain all its nutrition from that host. Unlike grazers or predators, parasites cannot move from host to host, supplementing their diet with a variety of foods.

Estimates of the number of species worldwide have risen sharply in recent decades owing to research revealing parasitic species to be much more specific to one host species than previously realized. What once may have been considered a single parasitic species attacking many different host species has often turned out to be a group of very similar, yet distinct, parasitic species, each specialized to its own host. This speciation occurs because different parasitic populations become adapted to living on different hosts and coping with the defenses of these hosts. Over time, many of these different parasite populations evolve into genetically distinct species. It is through the specialization of individuals of a species onto different hosts, ultimately resulting in speciation, that parasitism appears to have become the most common way of life on Earth.

For example, swallowtail butterflies (*Papilio*) include more than 500 species worldwide. In most species an adult female lays her eggs on a host plant, and, after they hatch, the caterpillars complete their development by feeding parasitically on that plant. In North America there are two groups of these butterflies that have evolved to use different hosts: the tiger swallowtail group and the Old World swallowtail group (*Papilio machaon*). In the Old World swallowtail group are several species that feed on plants in the carrot family Apiaceae (also called Umbelliferae), with different populations feeding on different plant species. However, one species within this group, the Oregon swallowtail (*Papilio oregonius*), has become specialized to feed on tarragon sagebrush (*Artemisia dracunculus*), which is in the plant family Asteracaea (Compositae of some sources). Among the tiger swallowtail group, various members have become specialized to different plant hosts. The eastern tiger swallowtail (*Papilio glaucus*) has a long list of recorded hosts, but it is now known that the northern and southern populations are adapted to different plant species, and these populations cannot develop on the others' hosts.

Although many parasitic species complete all developmental stages on a single host individual, thousands of other parasitic species alternate between two or more host species, specializing on a different host species at each developmental stage. Many parasites, from a diverse array of species such as certain viruses, flatworms, nematodes, and aphids, specialize on different host species at different stages of development. Among aphids alone at least 2,700 species alternate among hosts.

Parasites have evolved by three major evolutionary routes to alternate among two or more hosts. Some parasite species have evolved to alternate between their final host and an intermediate host, or vector, that transfers the parasite from one final host to another. For example, the malarial parasite *Plasmodium falciparum* alternates between a final human host and an intermediate mosquito host by which the parasite is transferred from one person to another. The parasite uses the mosquito as a mobile hypodermic syringe. Examples of a similar kind of transmission between a final host and an intermediate host with piercing mouthparts occur in many other species. Viruses, rickettsias, protozoa, and nematodes all have species that are transmitted between vertebrates through biting flies. Some viruses and other parasites are similarly transmitted between plant species by aphids, whose piercing mouthparts transmit the parasites directly into plant tissues while the aphids are feeding.

Other parasites alternate between a host and the predator that eats it. These parasites have turned an evolutionary problem (being killed along with their host) into an evolutionary opportunity (being transferred to the predator and continuing to feed). As it develops, the parasite attacks hosts higher in the food chain, alternating between herbivore and predator or between an intermediate and a top predator in the food chain. Many parasites alternate between snails or other invertebrates and vertebrate predators that feed upon these invertebrates, whereas others alternate among vertebrate species. The pork tapeworm (*Taenia solium*), for example, alternates between pigs and humans in societies in which improperly cooked pork is eaten.

Still other parasites employ wings, wind, or water to alternate between hosts. Many aphid species alternate between a summer host and a winter host by producing winged individuals that fly to the new host. Rust fungi such as wheat stem rust can be carried between hosts by wind currents, and the parasite *Schistosoma mansoni*, which causes the disease schistosomiasis, alternates between *Biomphalaria* snails and humans by moving through water.

The different ways by which host species are linked to parasites contribute to the complex web of interactions that shape the structure of communities. The effect of parasitism on the dynamics of populations and the organization of communities is still one of the most underexplored topics in ecology.

PARASITOID

A parasitoid is an insect whose larvae feed and develop within or on the bodies of other arthropods. Each parasitoid larva develops on a single individual and eventually kills that host. Most parasitoids are wasps, but some flies and a small number of beetles, moths, lacewings, and even one caddisfly species have evolved to be parasitoids. Parasitoids alone number about 68,000 named species, and most have yet to be named and described. Realistic estimates of the total number of described and undescribed parasitoid species are about 800,000.

Grazing

The word "grazing" conjures up images of large mammals moving through seas of grass. Grazing, however, is a form of interspecific interaction that has been adopted by a number of other groups as well. A grazer is defined as any species that moves from one victim to another, feeding on part of each victim without actually killing it outright. The "victim" is to the grazer as prey is to the predator. Hence, grasshoppers that jump from plant to plant, chewing a portion of the leaves of each one they visit, are grazers, as are caterpillars that crawl from one plant to another during development rather than remain as parasites on an individual plant. The grazing lifestyle differs from the parasitic lifestyle in a few important ways. Individuals can vary their diets with different foods, and, by not remaining attached to a single individual for long periods of time, their victims do not have time to develop induced specialized defenses, such as an immune response that a host can develop against a parasite.

Grazing is more commonly perpetrated on plants than animals because plants have a modular structure that allows a part of them to be lost without the whole individual being destroyed. In contrast, most animals that lose a part of the body to an antagonist die immediately or soon afterward, rendering the interaction an act of predation rather than grazing. An exception to this rule occurs in species that can disconnect body parts—some lizards and salamanders are able to detach their tails if they are attacked by a predator.

On most continents, reciprocal evolutionary changes, or coevolution, between grasses and large grazing mammals have taken place over periods of millions of years. Many grass species have evolved the ability to tolerate high levels of grazing, which is evident to anyone who regularly mows a lawn. Simultaneously, they have evolved other defenses, such as high silica content, which reduces their palatability to some grazers. A number of herbivorous mammals have responded to these defenses by evolving the ability to specialize on grasses with high silica content and low nutritional value. Many large grazing mammals such as elephants have high-crowned (hypsodont) teeth that are constantly replaced by growth from below as the crowns are worn down by the silica in their food. Many of these species also have complicated digestive systems with a gut full of microflora and microfauna capable of extracting many of the nutrients from the plants.

Not all grasslands, however, are adapted to grazing by large mammals. In North America, although the grasslands of the Great Plains coevolved with large herds of bison, the grasslands of the upper Intermontane West (which roughly includes eastern Washington and Oregon) have never supported these large grazing herds. The Great Plains had grasses that formed sods and could withstand trampling by large-hooved mammals. These sods were so tightly interwoven that early European settlers cut them to use as roofs for their houses. The grasses of the Intermontane West, however, were tuft grasses that did not form sods and quickly died when trampled. Consequently, when cows replaced bison as the large, grazing mammal of North America, the grasslands of the Great Plains sustained the grazing pressure, whereas those of the Intermontane West rapidly eroded. Similar problems have arisen in other parts of the world where cattle have been introduced into grasslands that did not have a history of coevolution with large grazing mammals.

Plants have evolved more than 10,000 chemical compounds that are not involved in primary metabolism, and most of these compounds are thought to have evolved as defenses against herbivores and pathogens. Some of these chemical compounds are defenses against grazers, whereas others are defenses against parasites. Most of the chemical compounds that make herbs so flavourful and useful in cooking probably evolved as defenses against enemies. These compounds, called allelochemicals, are found in almost all plant species, and their great diversity suggests that chemical defense against herbivores and pathogens has always been an important part of plant evolution.

Predation

Predation differs from both parasitism and grazing in that the victims are killed immediately. Predators therefore differ from parasites and grazers in their effects on the dynamics of populations and the organization of communities. As with parasitism and grazing, predation is an interaction that has arisen many times in many taxonomic groups worldwide. Bats that capture insects in flight, starfish that attack marine invertebrates, flies that attack other insects, and adult beetles that scavenge the ground for seeds are all examples of the predatory lifestyle. Cannibalism, in which individuals of the same species prey on one another, also has arisen many times and is common in some animal species. Some salamanders and toads have tadpoles that occur in two forms, one of which has a specialized head that allows it to cannibalize other tadpoles of the same species.

Because predators kill their prey immediately, natural selection favours the development of a variety of quick defenses against predators. In contrast, the hosts of parasites and the victims of grazers can respond in other ways. A parasitized host can induce defenses over a longer period of time as the parasite develops within it, and a plant population subjected to grazing can evolve traits that minimize the effects of losing leaves, branches, or flowers. Therefore, the evolution of interactions between parasites and hosts, grazers and victims, and predators and

D'Orbigny's round-eared bat (Tonatia silvicola) *capturing a katydid in flight.* © Merlin D. Tuttle, Bat Conservation International/Photo Researchers, Inc.

prey all differ from one another as a result of the ways in which the interaction affects the victim.

Most predators attack more than one prey species. Nevertheless, there are some ecological conditions that have permitted the evolution of highly specialized predators that attack only a few prey species. The evolution of specialization in predators (and in grazers) requires that the prey species be predictably available year after year as well as easy to find and abundant throughout the year or during the periods of time when other foods are scarce. In addition, the prey must require some form of specialization of the predator to be captured, handled, and digested successfully as the major part of a diet. The most specialized predators attack prey that fulfill these ecological conditions. Examples include anteaters, aardwolves, and numbats that eat only ants or termites, which are among the most abundant insects in many terrestrial communities. Among birds, snail kites (*Rostrhamus sociabilis*) are perhaps the most specialized predators. They feed almost exclusively on snails of the genus *Pomacea*, using their highly hooked bills to extract these abundant snails from their shells.

Some seed predators are also highly specialized to attack the seeds of only one or a few plant species. (Seed consumption is considered predation because the entire living embryo of a plant is destroyed.) Crossbills exhibit one of the most extreme examples of specialization. These birds have beaks that allow them to pry open the closed cones of conifers to extract the seeds. The exact shape of their bills varies among populations and species in both North America and Europe. Experiments on red crossbills (*Loxia curvirostra*) have shown that different populations of these birds have bill sizes and shapes that have been adapted to harvest efficiently only one conifer

The silky anteater (Cyclopes didactylus) *extends its long, narrow tongue to capture and ingest prey.* Gunter Ziesler/Bruce Coleman Ltd.

species. Hence, red crossbills are a complex of populations, each adapted to different conifer species.

Predators can greatly affect the structure of communities. For example, seed predators commonly scour the ground for each year's seed crop, eating most of the seeds produced by many different species each year. To defend against these predators, certain plants are thought to resort to mast seeding, which is the production of many seeds by the plant every two or more years in regional synchrony with other plants of the same species. Mast seeding is an effective defense because the seed predators become satiated before all the seeds have been consumed. The consequence of mast seeding for the organization of communities is that, instead of a few new seedlings establishing themselves every year, major pulses occur over time during which new plants become established and old plants die. Many conifers in boreal forests exhibit mast seeding as do other species such as bamboos. Some bamboo species grow for 100 years or more before producing seeds. Then all at once the bamboo plants over a large geographic region will set seed and die in the same year.

Competition

Competition is a powerful form of interaction in the organization of communities, but it differs from other forms of antagonistic and mutualistic relationships in that no species benefits from the interaction. In competitive interactions, species evolve either to avoid each other, to tolerate the presence of the other, or to aggressively exclude the other.

Species compete for almost every conceivable kind of resource, and the same two species may compete for different resources in different environments. Hole-nesting birds compete for tree holes, plant species compete for pollinators and seed dispersers, and male birds compete for preferred sites to defend as territories for attracting females. Species may compete for many resources simultaneously, but often one resource, called the limiting resource because it limits the population growth of each species, is the focus of competition. Moreover, the ways in which species compete vary with the resources. In some cases, species compete by capturing resources faster than their competitors (exploitation competition). Some plant species, for example, are able to extract water and nutrients from the soil faster than surrounding species. In other cases, the two species physically interfere with one another (interference competition) by aggressively attempting to exclude one another from particular habitats.

Over evolutionary time, the effects of competition on species can vary. In some environments, the effects may be highly asymmetrical, and, at the extreme called amensalism, the survival or growth of one species may be inhibited and the other(s) not affected. The weaker competitor will either go extinct locally, diverge from the other species in its use of resources, or evolve an increased competitive ability. All three outcomes have been observed in natural and experimental populations studied by ecologists.

Species diverge from one another through competition, with the result that they fill different niches within the community. The great differences in bill size and shape that some of Darwin's finches in the Galapagos have evolved have resulted from competition. This process, called character displacement, results as natural selection favours those individuals in each species that compete least with individuals of the other species. Experimental studies of coexisting seed-feeding rodents in the deserts of North America have shown that these species have evolved differences in size and other characteristics to minimize competition.

By evolving in response to one another, many competitors may be able to coexist regionally over the long term but not locally. Within any local area, one species may generally be driven to extinction by the other. Which species wins locally will depend on the physical environment, the genetic makeup of each of the competing species, and their interactions with other species in the community. Even subtle changes in the environment can affect which species wins. Experiments with species of flies (*Drosophila*) have shown that, when all other factors are held constant, small variations in temperature or in the percentage of ethanol in the larval environment can determine which species outcompetes the other. Hence, the continued coexistence of some competing species may depend critically on multiple populations of both or all species being distributed over a number of environments throughout a region.

Commensalism and Other Types of Interaction

In commensal interactions, one species benefits and the other is unaffected. The commensal organism may depend on its host for food, shelter, support, transport, or a combination of these.

One example of commensalism involves a small crab that lives inside an oyster's shell. The crab enters the shell as a larva and receives shelter while it grows. Once fully grown, however, it is unable to exit through the narrow opening of the two valves, and so it remains within the shell, snatching particles of food from the oyster but not harming its unwitting benefactor. Another form of commensalism occurs between small plants called epiphytes and the large tree branches on which they grow. Epiphytes depend on their hosts for structural support but do not derive nourishment from them or harm them in any way.

Many other kinds of interaction, however, range from antagonism to commensalism to mutualism, depending on the ecological circumstances. For example, plant-feeding insects may have large detrimental effects on plant survival or reproduction if they attack small or nonvigorous plants but may have little or no effect on large or vigorous plants of the same species. Some human diseases may cause only temporary discomfort or be life-threatening, depending on the age and physical condition of the person.

No interaction between species fits neatly into the categories of antagonism, commensalism, or mutualism. The interaction depends on the genetic makeup of both species and the age, size, and physical condition of the individuals. Interactions may even depend on the composition of the community in which the interaction takes place. For example, the moth *Greya politella* pollinates the flowers of a small herb called the prairie star (*Lithophragma parviflorum*). The female moth pollinates while she lays eggs (oviposits) in the corolla of the flower. As she pushes her abdomen down into a flower, pollen adheres to her. She flies on to the next flower to lay more eggs, where some of the pollen rubs off onto the stigma of the flower, causing pollination to occur. Although this unusual pollination mechanism is very effective in some local populations, in other communities different pollinators such as bee flies and bees are so common that their visits to the flowers swamp the pollination efforts of the moths. As a result, pollination by the moths makes up a very tiny proportion of all the pollinator visits that occur within that community and probably has little effect on plant reproduction or natural selection. This moth, therefore, is a commensal in some populations and a mutualist in others, depending on the local assemblage of pollinator species.

THE COEVOLUTIONARY PROCESS

As pairs or groups of species interact, they evolve in response to each other. These reciprocal evolutionary changes in interacting species are called coevolutionary processes, one of the primary methods by which biological communities are organized. Through coevolution local populations of interacting species become adapted to one another, sometimes even evolving into new species.

The Study of Coevolution

To understand how coevolution shapes interactions within communities, researchers must distinguish between traits that have coevolved and those that were already present in ancestors before the interspecific interaction began.

For example, hummingbirds use their wings and bills to reach the nectar within flowers. A hummingbird with a long bill may have evolved its bill as a result of coevolution with a particular species of flower; however, its wings are not the result of coevolution. Wings were already present in birds before hummingbirds evolved. Therefore, both the evolutionary ecology and the history (phylogeny) of the interacting species must be studied. The phylogeny indicates when each species arose within a lineage and when each new trait made its first appearance. The ecological studies can then show how each of those traits has been shaped by and used under different ecological conditions.

The study of the coevolution between moths of the family Prodoxidae and their host plants illustrates the interplay of phylogeny and ecology. Prodoxid moths include some species that have become major pollinators of plants. These pollinators include yucca moths (of the genera *Tegeticula* and *Parategeticula*) and *Greya* moths. *Greya* moths inadvertently, or passively, pollinate the flowers they visit, but their close relatives the yucca moths purposely, or actively, perform this function. Female yucca moths collect and carry pollen on specialized appendages attached to their mouthparts. They visit yucca flowers to lay their eggs in the floral ovary, and their offspring feed on the developing seeds. While visiting each flower, a female moth takes some of the pollen she is carrying and places it directly on the stigma of the flower. Her offspring are therefore guaranteed developing seeds on which to feed. The yuccas have evolved to depend solely on these moths for pollination. Unlike many other plant species, they do not produce nectar or any other reward for pollinators and so do not waste energy to attract pollinators. They lose some of their seeds to the yucca moth larvae, but this is the cost of coevolution with this highly efficient pollinator.

Phylogenetic studies have shown that the loss of nectar production in yuccas and the evolution of active pollination in yucca moths are novel traits that have arisen through coevolution, as the relatives of yuccas produce nectar, and the relatives of yucca moths, the *Greya* moths, do not actively pollinate their host plants. Some other aspects of the interaction make use of traits that did not coevolve between yuccas and yucca moths. Instead, the traits were present in ancestors. Laying eggs in flowers and local specialization to one plant species are two traits that are common to all the close relatives of yucca moths, regardless of the plants on which they feed.

Therefore, by combining ecological and phylogenetic studies, researchers can piece together the history of coevolution between these species. The coevolved mutualism between yuccas and yucca moths began when their ancestors inadvertently became more successful at survival and reproduction as a result of their interactions. Yuccas that did not waste energy on nectar production to attract other pollinators achieved an advantage over those plants that did. Those that ensured the availability of developing seeds for their offspring by actively pollinating the flowers in which they laid their eggs also gained an advantage over populations that did not do so. The process undoubtedly involved many other twists and turns along the way, but the combination of evolutionary ecological and phylogenetic studies allows at least part of the coevolutionary process to be reconstructed.

All coevolved interactions are similar to those between yuccas and yucca moths in that natural selection operates on traits that are already present within species, molding them in new ways by favouring new mutations that fine-tune the relationship. Tinkering rather than engineering is how the biologist François Jacob described the process of evolution, and his analogy certainly extends to the coevolutionary process. Coevolved interactions are not designed from scratch for maximum efficiency. Instead, evolution fiddles with existing structures and behaviours and adapts them to perform new functions—in effect, jury-rigging them. Consequently, the organization of biological communities reflects this makeshift nature of adaptation and coevolution.

The Coevolutionary "Arms Race" Versus Reduced Antagonism

Nothing is absolutely predictable about the direction of coevolution. How an interaction coevolves depends not only on the current genetic makeup of the species involved but also on new mutations that arise, the population characteristics of each species, and the community context in which the interaction takes places. Under some ecological conditions, an antagonistic interaction between two species can coevolve to enhance the antagonism. The species "build up" methods of defense and attack, much like an evolutionary arms race. Under other ecological conditions, however, the antagonism may be lessened, resulting in reduced antagonism.

In an evolutionary arms race, natural selection progressively escalates the defenses and counterdefenses of the species. The thick calcareous shells of many marine mollusks and the powerful drilling appendages and musculature of their predators are thought to have coevolved through this process of escalation. A similar example of coevolution has occurred in the endemic mollusks and crabs in Lake Tanganyika. The mollusks in this lake have much thicker shells than other freshwater mollusks, and the endemic crab that feeds on them has much larger chelae (pincerlike claws) than other freshwater crabs. Differences between these mollusks and crabs and the freshwater species throughout the world to which they are related appear to be caused by coevolution rather than any unique nutrient or mineral conditions in this lake.

Parasites and their hosts engage in a similar evolutionary arms race, although in the past parasitologists believed this not to be the case. Instead, parasites were thought to evolve gradually toward reduced antagonism—having a less detrimental effect on their hosts. The degree of virulence was sometimes regarded as an indicator of the age of the relationship: a particularly virulent relationship, which resulted in the swift demise of the host, was considered new. Research in population biology and evolutionary ecology, however, provided evidence that contradicts this view. Parasite-host interactions are now understood to evolve toward either increased or decreased antagonism, depending on several important ecological factors.

The density of the host population and the transmission rate of the parasite are two of the most important of these ecological factors. The density of the host species determines how often the opportunity arises for a parasite to move from one host to another. The parasite's transmission rate determines how easily a parasite can move between hosts when the opportunity does arise. Only some parasites are transmitted easily between hosts. If the host occurs at a high density and the transmission rate of the parasite is also high, natural selection favours increased virulence in the parasite. Being easily transmissible and having many host individuals to infect, the parasite can multiply quickly and escape to new hosts before it kills its current host and dies along with it. Some forms (genotypes) of the parasite will already contain or will develop mutations that increase the speed and proficiency of this process. By producing more organisms that survive, the mutated form of the parasite will outcompete those parasites with the original genotype that are not able to maximize the opportunity to infect the greatest number of hosts. After several generations, many more parasites with enhanced virulence will exist, and this genotype can be said to be favoured by natural selection. If host population density remains high, the parasite genotype that confers the most virulence will become the only form of the parasite in that population.

At the other extreme, if the host population density is low and the transmission rate of the parasite is also low, natural selection will favour less virulent forms of the parasite. The highly virulent forms that quickly kill their host will often die along with their host without having spread to other hosts, leaving only the less virulent parasites to propagate the species.

In many natural environments, host populations fluctuate between high and low density. Consequently, the parasite population will fluctuate as well, sometimes containing more highly virulent forms, sometimes less virulent forms. Depending on the rate at which host density fluctuates, the host population will vary in the degree and mix of virulent forms that it harbours.

The evolution of myxoma virus in rabbits in Australia shows how quickly coevolution of parasites and hosts can proceed to a new outcome, in this case intermediate virulence. European rabbits were introduced into Australia in the 1800s. In the absence of parasites and predators that had kept their numbers in check in their European habitat, they multiplied and disseminated rapidly, causing widespread destruction of the native vegetation. When the myxoma virus was introduced into Australia in 1950 to control rabbit populations, it was highly virulent and caused death in almost all infected rabbits within two weeks. Since then, however, coevolution of the virus and rabbit populations has occurred, resulting in an interaction less immediately lethal to the rabbits. As population levels of the rabbits decreased, mutant strains of the virus that allowed a rabbit host to live longer were favoured, thereby increasing the opportunity for the virus to spread to another rabbit before its current host died. Most infected rabbits still die from the infection, but death is not as relentless and most infected individuals survive for two and a half to four weeks after infection.

Coevolution and the Organization of Communities

The coevolution of the myxoma virus and rabbit species described above illustrates how this process operates to maintain the organization of biological communities, averting the havoc that might ensue without the proper checks and balances that the process ensures. Unfortunately, the importance of interspecific interactions may become apparent only after the balance of a community has been disrupted, often by human hands and often with serious reverberations. If the rabbit species had not been introduced into a community in which none of its natural predators were found, its potential for devastation would not be fully appreciated. Interspecies interactions are necessary to maintain population levels of moderate size, and they function in most biological communities in much the same way that the rabbit population was controlled by myxoma virus.

As biological communities are dismantled through human activities, coevolutionary processes and their effects on the organization of communities are disrupted. Changes in population density and the introduction of new species can cause the extinction of other species. In the process, the way natural selection acts on the remaining species within those communities is altered. Increased population densities of some species, for example, are likely to favour the evolution of more virulent genotypes of some parasites. Low population levels of other species can cause the chance loss of genes important to the ongoing coevolution of other interactions. It will seldom be possible to predict how these changes in coevolution will affect the organization of communities, because of the intricate interdependencies among all organisms of the biosphere.

Gene-for-Gene Coevolution

In some interactions between parasites and hosts, coevolution can take a specific form called gene-for-gene coevolution or matching-gene coevolution. It is a form of reciprocal evolutionary change based on the idea that, if one member of a coevolving relationship has a gene that affects the relationship, the other member has a gene to counter this effect. These genes evolve reciprocally and provide the genetic basis for certain types of coevolution. This relationship has been demonstrated between plants and a number of their parasites, including rust fungi, nematodes, bacteria, viruses, and one insect species. Its principles also form the basis of many plant breeding programs designed to increase resistance against pathogens.

The process of gene-for-gene coevolution begins when a parasite population encounters a new plant host. Most host individuals will not be able to detect the presence of the parasite. Certain host individuals, however, may have a mutated gene, dubbed the resistance gene in this scenario, that allows them to detect a substance the parasite emits, encoded by a so-called avirulence gene. After being alerted to the threat of the parasite, the host responds to prevent the parasite from invading. The resistance gene will confer an advantage to plants that carry it, allowing individuals to survive and pass on their genotype to future generations. Individuals that do not possess this gene will not be able to resist invasion by the parasite and will die, unable to pass on their genotype. Thus, the new resistance gene will spread through the plant population. At this point the parasite might seem to be outwitted, but actually it may be able to circumvent this genetic evasion by the host with a genetic trick of its own. If a mutation arises in the gene that codes for the product that the host recognizes, the gene product will be altered and the host will no longer be able to resist the parasite. The spread of this mutant gene in the parasite population will be favoured by natural selection. A genetic Ping-Pong match between the two species can then ensue, as the host develops another mutation in any gene that allows it to detect the parasite, and the parasite responds to this defensive maneuver with a genetic alteration to avoid detection. The host and parasite populations therefore coevolve by the accumulation of these matching genes.

In agriculture, gene-for-gene relationships are maintained by introducing new resistance genes into all plants that cover a large area. In natural populations, each new resistance gene appears as a mutant in a single individual and then spreads by natural selection throughout the population in subsequent generations. Demonstrating a gene-for-gene relationship in natural populations is a difficult and time-consuming process because it demands detailed genetic and ecological studies of the plants and their pathogens that take many years.

The best-studied example is that of wild flax (*Linum marginale*) and flax rust (*Melampsora lini*) in Australia. Local populations of flax plants and flax rust harbour multiple matching genes for resistance and avirulence. The number of genes and their frequency within local populations fluctuate greatly over time as coevolution continues. In small populations, the resistance genes can be lost by chance alone through the process of genetic drift. New genes in the host and parasite populations can appear through either mutation or the influx of genes from other populations. Consequently, the long-term dynamics of the gene-for-gene coevolution between flax and flax rust depend on the rate at which new genes appear within local populations of the parasite and host, the intensity with which natural selection acts on these genes (which, in turn, depends on the virulence of the particular parasite genotype), the population sizes of both host and parasite, and the rate that genes are transferred among populations.

Not all interactions between plants and parasites coevolve in a gene-for-gene manner. Resistance in a plant host is often determined by many genes rather than by a single gene. Examples of gene-for-gene coevolution, however, are slowly accumulating, and these are providing a powerful tool in breeding crop plants that are resistant to pathogens and parasites. As other forms of coevolution are studied in natural populations, the results will help determine still other ways of selecting for more durable resistance in crop plants. Such studies, however, require that intact biological communities are preserved as precious natural laboratories for understanding the coevolutionary process.

The Geographic Mosaic Theory of Coevolution

The study of evolving interactions in natural biological communities has indicated that the long-term dynamics of coevolution may occur over large geographic ranges rather than within local populations. This view is called the geographic mosaic theory of coevolution. It is based on the observation that a species may adapt and become specialized to another species differently in separate regions. A species that is involved in an interspecific interaction in one geographic area may not even be present in another geographic area. This geographic mosaic in evolving interactions provides the raw material for the overall direction of coevolution, which proceeds as genes that are favoured in local interactions spread out into other populations.

Some local populations may contribute little to the overall direction of coevolution between two or more species, whereas other populations may be crucial to the process. A highly virulent form of a parasite that recently has been introduced into a local population may cause the extinction of its host population, thereby causing its own extinction. In another population the evolution of a powerful host resistance gene may cause the local extinction of the parasite population. In still other populations the two species may continue to coexist but coevolve in different ways. Over a long period some adaptations in these populations will spread to other populations and influence the overall direction of coevolution between the parasite and host species. Similar geographic differences in interactions are known to occur between predators and prey and between competitors and mutualists.

For some forms of coevolution to occur at a geographic level, many populations of the interacting species must be maintained on a local scale (metapopulations) as well as across broader geographic ranges. Should a species be reduced to a few populations, the geographic mosaic of diverse adaptations that fuels the coevolutionary process is diminished.

Coevolution of One Species with Several Species

In the process called coevolutionary alternation, one species coevolves with several other species by shifting among the species with which it interacts over many generations. European cuckoos (*Cuculus canorus*) provide an example of this type of coevolution. The cuckoos behave as brood parasites, laying their eggs in the nests of other avian species and depending on these hosts to raise their young. The four major host species for cuckoos in Britain are meadow pipits (*Anthus pratensis*), reed warblers (*Acrocephalus scirpaceus*), pied wagtails (*Motacilla alba yarrellii*), and dunnocks (*Prunella modularis*).

Cuckoo populations have evolved many adaptations that enable them to trick their hosts into rearing their young, the most impressive of which is the production of eggs that resemble those of their host. Cuckoos can produce eggs that are similar in colour to the eggs of meadow pipits, reed warblers, or pied wagtails. Three different cuckoo genotypes are responsible for the production of these three different egg colours, and thus cuckoos are said to be polymorphic with respect to egg colour. These genotypes are maintained in cuckoo populations because natural selection is continually changing which genotype is favoured. This occurs because after many generations the host species can develop defenses against the cuckoo, such as the ability to discriminate between eggs and eject those that have only minute differences from their own. (That this ability to reject eggs is evolved is evidenced by the fact that in Iceland, where cuckoos are not indigenous, pied wagtails and meadow pipits will accept cuckoo eggs placed in their nests by researchers.) As the gene for this defensive maneuver spreads throughout the host

The reed warbler (Acrocephalus scirpaceus) *acts as one of several host species for the cuckoo in coevolutionary alternation.* Stephen Dalton/Natural History Photographic Agency

population, cuckoos with genotypes that allow them to produce eggs mimicking the egg colour of a new host are favoured.

By alternating among hosts, cuckoos are coevolving with a number of bird species in Britain. Some of their current hosts, including dunnocks, have few defenses against cuckoos and do not reject cuckoo eggs, which may indicate that these hosts are relatively recent targets. Some other potential hosts that cuckoos do not currently colonize will strongly reject any eggs that are not exactly

Coevolution of one species with many species. (Left) *European cuckoo* (Cuculus canorus) *egg in the nest of a reed warbler* (Acrocephalus scirpaceus) *and two smaller reed warbler eggs.* (Right) *Fledgling cuckoo, having kicked the young reed warbler chicks from the nest, being fed by an adult reed warbler, which accepts the chick as her own.* John Markham/Bruce Coleman Ltd.John Markham/Bruce Coleman Ltd.

like their own, suggesting that these bird species may have been hosts in the recent past but were abandoned by cuckoos after defenses against them evolved. Over many generations, these former hosts will probably lose their defenses because the evolutionary pressure to retain them has been relaxed—they are no longer being attacked by cuckoos. In Britain the loss of these kinds of defenses makes the birds once again targets for attack by cuckoos, thereby continuing the process of coevolutionary alternation among hosts.

Coevolution Among Groups of Species

Coevolution often involves even larger numbers of species, but many of these coevolving interactions are much more difficult to study than paired or alternating relationships. Examples of these larger groups of coevolving species include many butterflies that have coevolved with plants that produce showy flowers to attract them and fruit-eating birds that have coevolved with plants that produce small, fleshy fruits to disperse their seeds. Many of these interactions may have begun with a relationship between only a few species of animals and plants, but other species have evolved convergently, developing similar characteristics to exploit the existing relationships.

Convergence

Unrelated species living in similar physical environments often are shaped by natural selection to have comparable morphological, physiological, or life history characteristics, and they are said to evolve convergently. Convergence is a common feature of evolution and has major effects on the organization of biological communities. Interactions as well as characteristics can converge. Once an interaction evolves between two species, other species within the community may develop traits akin to those integral to the interaction, whereby the new species enters into the interaction. This type of convergence of species has occurred commonly in the evolution of mutualistic interactions, including those between pollinators and plants and those between vertebrates and fruits: some species drawn into the interaction become comutualists, contributing as well as benefiting from the relationship, whereas others become cheaters that only exploit the relationship. Either way, these additional species may influence the future evolution of the interaction.

A clear example of this kind of convergence of species is that between flowers and hawkmoths. In the tropical dry forest of Cañas in northwestern Costa Rica, there are 65 hawkmoth species and 31 native plant species adapted for hawkmoth pollination. The hawkmoth species are all members of one moth family called the Sphingidae. They have diverged into many species from a common moth ancestor, and it is therefore not surprising that they share the same basic hawkmoth body plan. The plants adapted for hawkmoth pollination, however, are distributed throughout 14 plant families. These species have evolved convergently from different ancestors to have floral shapes that attract hawkmoths.

Mimicry Complexes

A different kind of convergence has occurred in the evolution of mimetic butterflies and other insects. Mimicry occurs when two or more species evolve to resemble and sometimes behave in ways similar to another species. The most famous examples of mimicry are found among insects, and they take two forms: Müllerian mimicry, in which two species evolve convergently to have a similar appearance, and Batesian mimicry, in which one species evolves to resemble another. These different forms of mimicry are named after their 19th-century discoverers, the naturalists Fritz Müller and Henry Walter Bates. In the several decades following the publication of Charles Darwin's *On the Origin of Species* in 1859, mimicry was the major example used to show how evolution occurs through the mechanism of natural selection.

Müllerian mimicry can occur between two species that are distasteful to the same predators. Their predators learn to recognize and avoid distasteful prey by signals such as the colour patterns of wings. If two distasteful species develop the same colour pattern, the predator has to learn only one pattern to avoid, speeding up the learning process and providing an advantage to the convergent prey species. One of the distasteful species may initially model itself on the other, but, if they are almost equal in abundance, the species may coevolve and converge on some intermediate pattern. *Heliconius* butterflies in Central and South America form mimicry complexes of two or more species, and the colour patterns that result from this convergence vary geographically.

In Batesian mimicry a palatable species models itself on an unpalatable species to fool predators into believing that they are not tasty. Many flies have evolved to mimic bees, and some palatable butterflies have evolved to mimic unpalatable butterflies. If the mimic is uncommon, the convergence may not affect the unpalatable model, because it will be less likely that predators will consume many mimics by mistake and uncover the fraud. If the mimic, however, is abundant, its predators may eventually learn to dissociate its colour pattern with distastefulness because enough mimics would be inadvertently consumed and found palatable. Natural selection eventually would favour the evolution of a new colour pattern in the model species.

Coevolution Among Many Species

Coevolution between birds and fruit-bearing plant species is even more complicated than that between flowers and pollinators or between models and mimics, because so many plant species have evolved fleshy fruits for dispersal by birds and so many bird species have become adapted to eat fruits as part of their diets. Almost half of the 281 known terrestrial families of flowering plants include some species with fleshy fruits. About one-third of the 135 terrestrial bird families and one-fifth of the 107 terrestrial mammal families include some partly or wholly frugivorous species. Moreover, the evolution of these interactions is not limited to relationships between species within local communities. Many frugivorous birds migrate thousands of miles every year, and the ripening of the fruits of many plant species in temperate regions appears to be timed to the peak of bird migrations in the autumn. Consequently, the evolution of interactions between birds and fruits occurs over very broad geographic ranges. These interactions link more species in more communities than any other form of relationship among species. They show that the conservation of species demands a geographic, even global, perspective on how interactions between species are maintained within biological communities.

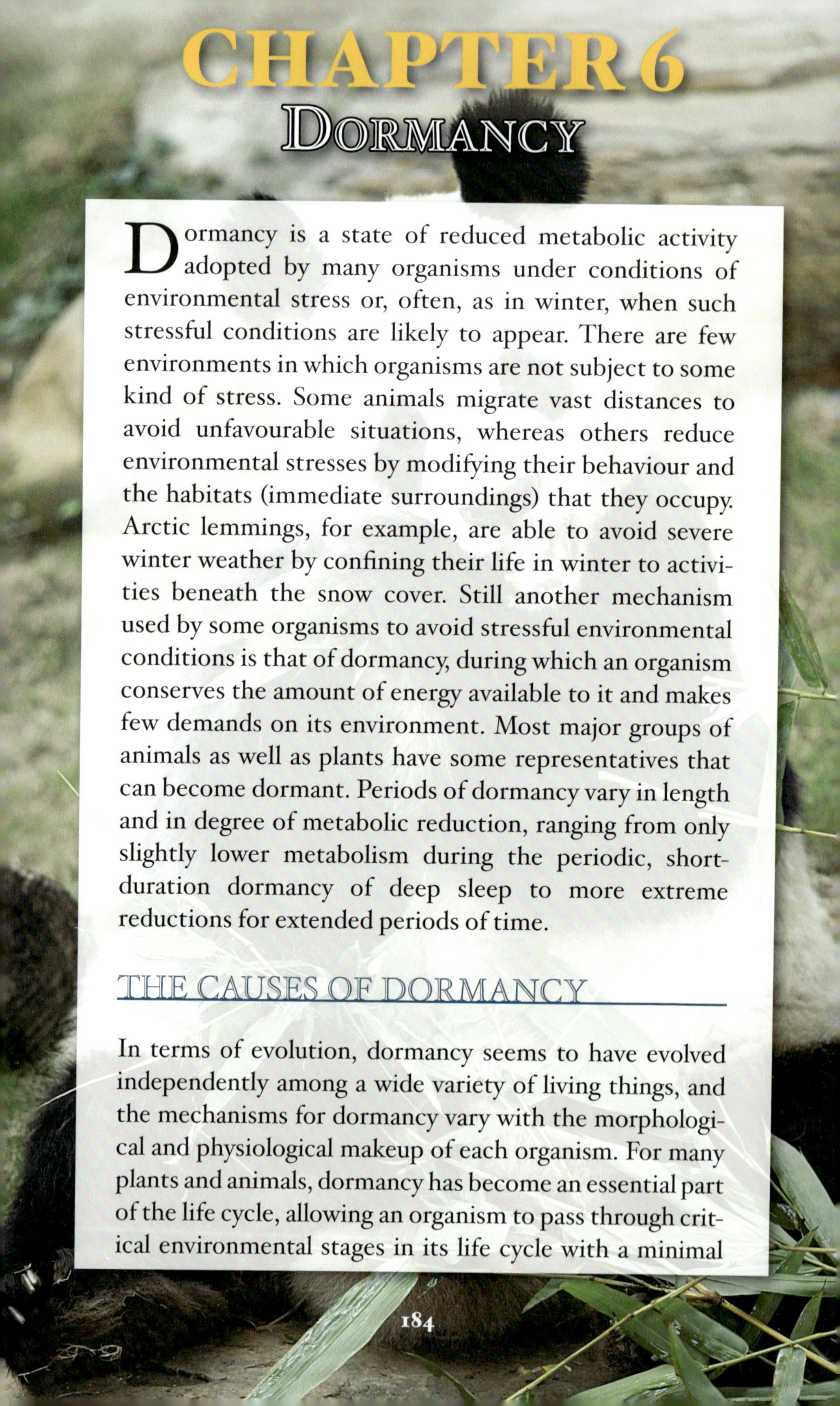

CHAPTER 6

Dormancy

Dormancy is a state of reduced metabolic activity adopted by many organisms under conditions of environmental stress or, often, as in winter, when such stressful conditions are likely to appear. There are few environments in which organisms are not subject to some kind of stress. Some animals migrate vast distances to avoid unfavourable situations, whereas others reduce environmental stresses by modifying their behaviour and the habitats (immediate surroundings) that they occupy. Arctic lemmings, for example, are able to avoid severe winter weather by confining their life in winter to activities beneath the snow cover. Still another mechanism used by some organisms to avoid stressful environmental conditions is that of dormancy, during which an organism conserves the amount of energy available to it and makes few demands on its environment. Most major groups of animals as well as plants have some representatives that can become dormant. Periods of dormancy vary in length and in degree of metabolic reduction, ranging from only slightly lower metabolism during the periodic, short-duration dormancy of deep sleep to more extreme reductions for extended periods of time.

THE CAUSES OF DORMANCY

In terms of evolution, dormancy seems to have evolved independently among a wide variety of living things, and the mechanisms for dormancy vary with the morphological and physiological makeup of each organism. For many plants and animals, dormancy has become an essential part of the life cycle, allowing an organism to pass through critical environmental stages in its life cycle with a minimal impact on the organism itself. When lakes, ponds, or rivers dry up, for example, aquatic organisms that can enter a period of dormancy survive, while others perish. Moreover, animals that can become dormant during the extreme cold of winter can extend their ranges into regions where animals incapable of dormancy cannot live. Dormancy also ensures that these animals will be free from competition during their periods of activity. Thus, dormancy is an adaptive mechanism that allows an organism to meet environmental stresses and to take advantage of environmental niches that otherwise would be untenable at certain times.

The dormant state that is induced in an organism during periods of environmental stress may be caused by a number of variables. Those of major importance in contributing to the onset of dormancy include changes in temperature and photoperiod and the availability of food, water, oxygen, and carbon dioxide. In general, because organisms normally exist within a relatively narrow temperature range, temperatures above or below the limits of this range can induce dormancy in certain organisms. Temperature changes also affect such other environmental parameters as the availability of food, water, and oxygen, thus providing further stimuli for dormancy. In Arctic regions, for example, certain animals become dormant during the winter months, when food is less abundant. In desert biomes, however, the summer months, which may be periods of reduced food availability, intense heat, or extreme aridity, stimulate some desert organisms to become dormant. The lack of water during summer periods of drought or winter periods of freezing, as well as annual changes in the duration and intensity of light, particularly at high latitudes, are other environmental factors that can induce dormant states.

Under natural conditions, most of the environmental variables that influence dormancy are interrelated in a cyclical pattern that is either circadian or annual. Fluctuations in the major daily variables—light and temperature—can induce rhythmical changes in the metabolic activity of an organism. Annual fluctuations in temperature and photoperiod can influence the availability of food and water. Concentrations of oxygen and carbon dioxide normally do not vary on a cyclical basis but as a result of habitat selection, such as burrowing in the mud, seeking a den, or other similar activities, in which the metabolic responses of the organism can alter the oxygen and carbon dioxide concentrations in its environment.

In an attempt to determine the relative influence of environmental factors upon dormancy, they have been varied experimentally. Investigations indicate that an organism, after it has adapted to a sequence of cyclical rhythms, tends to maintain its adaptive behaviour even though the environmental stimulus that originally elicited such behaviour is no longer present. For example, the Arctic ground squirrel (whose winter period of dormancy is referred to as hibernation), when taken into the laboratory, supplied with adequate amounts of food and water, and exposed to constant temperature and light, exhibits periodic torpor (extreme sluggishness)—an innate behavioral pattern that operates independently of environmental cues. Other animals frequently will continue to respond as if they were exposed to the cyclical changes of their home environments after they are removed from their natural habitats.

DORMANCY IN PROTOZOANS AND INVERTEBRATES

Protozoans and many invertebrates employ cysts and cystlike structures to protect themselves from temperature and other environmental extremes during their dormant states. These animals undergo a special "cyst" stage in their life cycle. In contrast, many insects can undergo diapause, a state characterized by the cessation of growth, throughout their life cycle.

Cysts and Cystlike Structures

Many parasitic and free-living protozoans (one-celled animals) exhibit a dormant stage by secreting a protective cyst. The stimulus for cyst formation in free-living protozoans may be temperature changes, pollution, or lack of food or water. *Euglena*, a protozoan that encysts to avoid environmental extremes, has two kinds of cysts. Apparently, one is formed only to avoid stressful conditions. The other is formed for the same reason but also involves asexual reproduction, resulting in a cyst that may contain up to 32 daughter organisms, which emerge under proper environmental conditions.

Free-living protozoans form cysts around themselves and avoid environmental extremes, but cysts are a part of the life cycle of parasitic protozoans. The causative agent of amebic dysentery, *Entamoeba histolytica*, is found in the intestine of infected individuals, in whom it forms cysts that pass to the outside in feces. When food or water containing cysts enters the digestive tract of another person, the amoebas are released from the cysts and infect the new host. Without encystment, which allows the organism to live in a dormant state in an unfavourable environment (e.g., water), amebic dysentery could be much more easily controlled. Protected by the cyst wall, however, the dormant contents of the cyst can survive for weeks. Although they are not particularly resistant to drying, the cysts of *E. histolytica* can withstand temperatures of up to 68 °C (154 °F) for five minutes. They are also resistant to certain chemicals.

Dormant cysts are formed during the life cycles of invertebrate parasites such as the oriental liver fluke (*Clonorchis sinensis*). The cyst stage of this organism develops in fish muscle; if the fish is eaten raw or undercooked, the encysted fluke is transferred to a new host. The encysted stage of the trichina worm (*Trichinella spiralis*), which causes trichinosis, is found in the muscle cells of hogs and is also an invertebrate parasite in which the dormant stage is an essential part of the life cycle. When undercooked pork is eaten, the cyst wall is dissolved by digestive juices, and the worm is able to make its way into the tissues of a new host.

The cystlike forms found in many other invertebrate groups are all dormant stages that preserve the species during times of environmental stress. All freshwater sponges and some marine species survive cold or drought by forming gemmules within the body of the adult sponge. These structures, which are surrounded by a resistant covering, are released when the sponge dies and disintegrates. When conditions are appropriate, the cell mass escapes from the covering and forms a new sponge.

Rotifers are microscopic aquatic animals that produce winter eggs with thick and resistant coverings similar to protozoan cysts. The eggs may remain dormant for long periods. They can survive drought or freezing and may be dispersed by wind or carried by animals. Thus, the cyst serves not only for survival of the egg under adverse conditions but also for dispersal. Some freshwater bryozoans develop disklike buds, or statoblasts, that are surrounded by a hard, chitinous (horny) shell. These statoblasts are the dormant structures that survive when the bryozoan dies in the fall or during a drought, forming a new bryozoan colony when favourable environmental conditions again prevail.

Among mollusks, land snails remain largely dormant throughout the day, with the soft head and foot withdrawn into the shell. During periods of drought or cold, they retreat into their shells and secrete a membrane (the epiphragm) of mucus and lime that covers the opening of the shell and resists desiccation. Slugs, on the other hand, bore into the ground and secrete a mucus mantle around themselves for protection during periods of unfavourable environmental conditions. Among the arthropods, many freshwater forms develop dormant cystlike stages that resist desiccation and allow the species to survive unfavourable periods.

Diapause in Insects

Many insects undergo periods of reduced metabolic activity called diapause. Diapause, which may occur during any stage of the life cycle—egg, nymph, larva, pupa, or adult—is usually characterized by a cessation of growth in the immature stages and a cessation of sexual activity in adults. In some insects, it is a reaction to unfavourable environmental conditions, whereas in others, such as certain moths and butterflies, diapause is a necessary stage of the life cycle. The 17-year larval and pupal periods of the cicada are examples of diapause. This form of dormancy is particularly common among insects that live in arid desert areas, where during the dry and hot summers, the insects usually hide themselves in the soil at suitable depths or under any available protective objects.

Insects may overwinter as egg, larva, nymph, pupa, or adult; because they can stand very low temperatures, few of these forms die if the winter temperatures are within their normal range. Even rather fragile forms, such as mosquitoes and butterflies, survive in sheltered, relatively dry places out of doors. Some butterflies even survive the winter in low shrubbery, where they may be completely covered by snow and ice for three or four months. Other insects prepare for winter by constructing nests or cocoons, while still others seek suitable hiding places.

Among some insect species, diapause lasts only until favourable environmental conditions return, after which the insect resumes its normal activities. In other species, favourable environmental conditions alone do not break the diapause; some other stimulus, such as cold or food, is necessary. The eggs of the mosquito *Aedes vexans*, for example, remain in diapause until the damp soil on which the eggs are laid is flooded to form a pool suitable for the larvae. The eggs of another mosquito, *Aedes canadensis*, are laid in the same soil as those of *Aedes vexans*, but they will not hatch until they have been subjected to cold. Thus, when both species lay their eggs together in early summer, those of *Aedes vexans* hatch in pools formed by late summer rains, but those of *Aedes canadensis* overwinter and hatch in the spring rain pools. Certain conditions are required to break diapause, and in some species (e.g., certain cutworms) a specific length of time must elapse before the stimuli are effective.

The onset of diapause depends on a combination of environmental factors operating on the regulatory mechanisms (i.e., nervous and endocrine systems) of the insect. Photoperiod and temperature influence the endocrine function of the brain, which synthesizes and secretes a substance (hormone) that controls other endocrine organs, specifically the prothoracic glands. Under the stimulation of the brain hormone, the prothoracic glands secrete a hormone called ecdysone. When stimulation by the brain hormone ceases, ecdysone is no longer secreted, and, in its absence, all insect growth and metamorphosis are halted. Thus, provision is made for the overwintering of immature insects in a state of developmental standstill. With the arrival of more favourable conditions, ecdysone is again secreted, and development resumes. Because many insect species have more than one generation of progeny per year, the prothoracic glands do not cease functioning except at some stage in the life cycle of the brood that must overwinter.

DORMANCY IN COLD-BLOODED VERTEBRATES

Two kinds of dormancy can be distinguished in vertebrates on the basis of body temperature. Most vertebrates are poikilothermous, or cold-blooded, because the body temperature follows that of the environment and is not kept constant by internal (homoiostatic) mechanisms. The second group, the homoiotherms, maintain a constant body temperature regardless of the ambient temperature, and includes the birds and mammals.

Fishes and Amphibians

The metabolism of poikilothermous animals is most influenced by the environmental variables of temperature, nutrition, and photoperiod. Photoperiod, the daily length of light exposure, has a marked metabolic effect in both fishes and amphibians. Fishes remain active throughout the year. However, their activity may be limited by temperature, as in those fish that rest on the bottom or in mud during cold periods. Brief superficial freezing and supercooling (without freezing) to temperatures below the freezing point of body fluids are experienced by resistant species, but it has not been established that fishes that have been frozen solid can become active when thawed. In the Arctic, no fishes are found in lakes that freeze solid in the winter.

Because most fishes do maintain some kind of activity year round, they cannot be said to become dormant in the sense in which the word is used in this section.

In addition to light and temperature, another environmental stress imposed upon fish is drought. Lungfishes, as represented by the African lungfish (*Protopterus*), burrow deeply into the mud when their water supply is diminished. They surround themselves with a cocoon of slime and remain inactive. Their gills are nonfunctional during this period of dormancy, and they use a lunglike air bladder for respiratory purposes. They rely on fat reserves as an energy source and excrete urea rather than ammonia to conserve water. Because ammonia as an excretory product is highly toxic, animals that excrete ammonia require large quantities of water to dilute it below toxic levels. Urea is a semi-solid substance of low solubility, and requires little or no water for its excretion. (Desert animals and many insects excrete urea.)

During periods of drought or cold, amphibians seek protective niches in which to remain dormant until the return of favourable environmental conditions. Overwintering of frogs and salamanders frequently involves their aggregation in large numbers in a moist terrestrial niche, such as a rotting log, in the mud on banks or bottoms of marshes and ponds, or in springs. The more terrestrially oriented amphibians, such as toads, may pass the winter in solitary burrows on land. During dry seasons, frogs may be dormant in a mud cocoon.

Reptiles

Because reptiles depend on external sources of heat to keep warm, they survive during periods of low temperature by seeking a place where the temperature will not fall below

SWIM BLADDER

A swim bladder, or air bladder, is a buoyancy organ possessed by most bony fish. The swim bladder is located in the body cavity and is derived from an outpocketing of the digestive tube. It contains gas (usually oxygen) and functions as a hydrostatic, or ballast, organ, enabling the fish to maintain its depth without floating upward or sinking. It also serves as a resonating chamber to produce or receive sound. In some species the swim bladder contains oil instead of gas. In certain primitive fish it functions as a lung or respiratory aid instead of a hydrostatic organ. The swim bladder is missing in some bottom-dwelling and deep-sea bony fish (teleosts) and in all cartilaginous fish (sharks, skates, and rays).

freezing, except temporarily. The most common niche for reptilian dormancy is almost always found underground at a depth dependent on the thermal conductivity of the soil relative to the minimum temperature reached. This factor alone can control the distribution of reptiles. None can survive in the Arctic or Antarctic in places in which the subsoil is permanently frozen. Relatively few can exist in areas near these regions, even if suitable sites for dormancy were available, because the short summers would prevent the completion of life cycles. Although the distribution of snakes at high latitudes or altitudes is limited, the adder has been found at 3,300 metres (10,000 feet) in the Swiss Alps and as far north as the Arctic Circle. The Himalayan pit viper has been found at an altitude of 5,000 metres (16,000 feet).

Dormancy is a state of torpor directly induced by low temperature. Reptiles may display a circadian rhythm, a seasonal one, or both. When the adder, for example, experiences temperatures of about 8–10 °C (46–50 °F), it begins to search out suitable niches in which to rest. Its dormancy ends on the first sunny days after the maximum temperature has reached 7.5 °C (45.5 °F). Because these conditions vary, the adder's period of dormancy extends from 275 days in northern Europe to 105 days in southern Europe and is about two weeks in the United Kingdom, where the Gulf Stream provides warmth.

Reptiles also normally become dormant during the hottest parts of summer, but the physiology of summer dormancy is quite different from that of winter. As already mentioned, winter dormancy is a state of torpor, induced by a low temperature, that becomes more pronounced as the temperature falls. There is, however, a wide range between the animal's normal, active (coenothermic) temperature and the lowest temperature at which it can exist. High temperatures, however, result in a much narrower range between the coenothermic temperature and temperatures that cause death. In other words, reptiles can tolerate colder temperatures much better than they can tolerate higher ones. For this reason, during hot weather they must seek refuge underground or in cool, shady places, where they remain physiologically active but must forego all normal activity because of the restricted nature of the cooler niche. Desert reptiles, in particular, exhibit such temperature responses daily.

During its dormancy, the amount of water needed by a reptile is less than at other times and is normally supplied by water produced from the metabolism of the animal's own stored food reserves, particularly fat. In areas where alternating wet and dry seasons occur, reptiles maintain a longer period of dormancy during the dry season. This behaviour may be related more to the lack of available water than to temperature, because in such areas the onset of the seasonal monsoons elicits a period of increased reptile activity.

MUD TURTLE

Mud turtles (*Kinosternon*) are any of about 18 species of semiterrestrial freshwater turtles belonging to the family Kinosternidae. Mud turtles are found in North and South America from New England to northern Argentina. Like the related musk turtles (*Sternotherus*), they are small animals (usually 15 cm [6 inches] or less in shell length) with fleshy barbels on the chin and the ability to exude a strong, musky odour. They differ from musk turtles by having a broad lower shell with a hinged section at either end. The hinged portions of the shells of some species can be pulled up to cover and protect the head, legs, and tail of the turtle.

Mud turtles occur in a variety of habitats that range from clear forest brooks to intermittent desert streams and ponds. Although highly aquatic in many respects—feeding is almost exclusively so—they are generally poor swimmers and instead prefer to walk along the bottoms of ponds and streams. In addition, some species, such as the striped mud turtle (*K. baurii*), survive drought periods through estivation (dormancy) under a shallow layer of mud.

Mud turtles are omnivores with a preference for animal matter, such as arthropods, worms, small fishes, and fish eggs. The clutch sizes of mud turtles vary from a single egg to nearly a dozen, depending on the female's size and health.

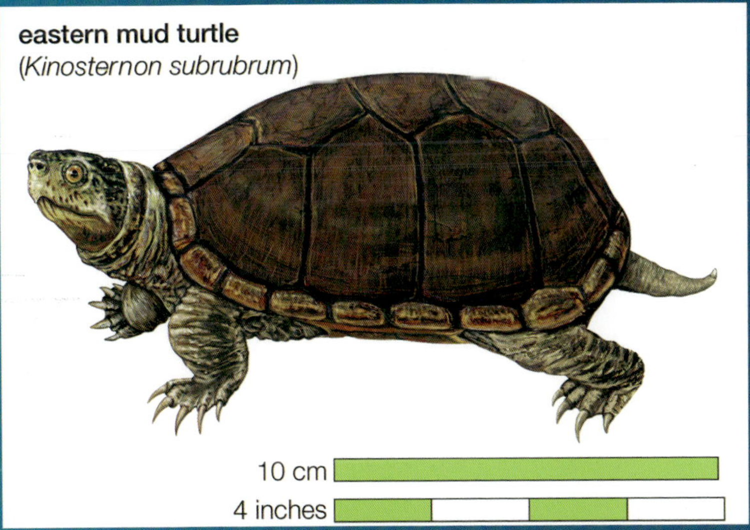

Mud turtles weather periods of drought through estivation (dormancy) beneath a thin mud layer. Encyclopædia Britannica, Inc.

Because there is only a limited number of suitable sites available for dormancy, several snakes, usually of the same species, may be found in each niche. As many as 100 or more snakes have been taken from one winter den. Occasionally, lizards and toads may also be found in the same den, but stories of snakes that share denning sites with small birds and mammals have been difficult to substantiate. It is much more commonplace to find that the entry of snakes into the burrow of a prairie dog or some other warm-blooded animal is followed by the evacuation of the original occupant.

Changes in latitude not only alter the lengths of the dormant and active periods of reptiles but also affect their circadian rhythms because of the changes in the proportions of night to day. Many species of snakes, including the adder, are normally active in the early evening. In the northerly latitudes (e.g., northern Europe, such as Scandinavia and Finland), where the length of the active season is reduced by as much as two-thirds, these snakes are active throughout the day and are able to take advantage of every warm hour to complete the necessary portions of their life cycle. Even this increased activity during the shorter summer season, however, does not compensate for the latitude. Growth and development slow to such a point that sexual maturity is delayed. The reproductive period requires two years rather than one, so young are produced only every other year instead of every year, as at lower latitudes.

DORMANCY, HIBERNATION, AND ESTIVATION IN WARM-BLOODED VERTEBRATES

The term hibernation is often loosely used to denote any state of torpor, inactivity, or dormancy that an organism might exhibit. Properly speaking, however, use of the term should be confined solely to warm-blooded homoiotherms (i.e., birds and mammals whose feathers or fur serve as insulation to reduce heat radiating from the body and aid in the maintenance of constant body temperatures, which normally are independent of those of the environment). Because warm-bloodedness gives animals an internal physiological stability, they are less dependent on many environmental restrictions, particularly those limitations imposed on organisms by ambient temperatures. For example, only two species of reptiles are found north of the Arctic Circle, but great numbers of birds live and breed there. Warm-bloodedness also signifies a high metabolic rate, a factor that undoubtedly influences normal learning, which depends heavily on the frequency and recency of experiences. Because periods of lowered metabolism interrupt continuous learning experiences, they may explain in part why birds and mammals are so much easier to train than any other animal. The benefits of warm-bloodedness require the expenditure of large amounts of energy through the year and make a heavy demand on available food supplies.

The term hibernation is also used to delineate the dormant state only during winter. In arid regions a reverse phenomenon called estivation is seen in which the animal becomes torpid during the hot, dry, barren summer. As a means of avoiding environmental stresses, hibernation and estivation are not common devices among warm-blooded animals, and they are far less common among birds than among mammals.

Some warm-blooded organisms exhibit thermic instability, a heterothermous condition that allows their metabolic rate to be reduced, with a commensurate reduction in body temperature. Heterothermy is a transitional state between cold-bloodedness and warm-bloodedness during which the animal is awake and moving. The body temperature, although not as constant as in humans, is not so low as to force the organism into deep hibernation. Among mammals, two monotremes, the spiny anteater and the duckbill platypus, are thermally unstable. Many marsupials, including the opossum, the pouched mouse, and the native cat (a weasel-like spotted marsupial of the family Dasyuridae), are also unable to maintain a fixed body temperature.

The true hibernator not only possesses adaptations that enable it to respond as a homoiothermous animal during certain periods of the year but can also adapt to stressful environmental situations and become essentially a poikilothermous animal during other periods. An animal exposed to food shortages, low temperatures, or lack of water, for example, may "turn off its thermostat" and hibernate until the environment becomes more favourable. Unlike poikilotherms, however, hibernators still retain a measure of temperature control and can change their metabolic levels as required. They can arouse themselves to full activity, whatever the environmental temperature, whereas the arousal of a poikilotherm is dependent upon increased environmental temperatures.

During the period prior to hibernation an animal must make a considerable number of gradual physiological and metabolic adjustments that appear to be correlated with temperature, light, and the availability of food. No one set of conditions applies equally to all hibernators: some store food, others do not; some become excessively fat, others gain a more moderate amount of weight. Generally, as the season advances and as the hibernator becomes progressively more prepared for hibernation, there is an increase of fat deposition and a general readjustment of body temperature, metabolism, and heart rate to lowered levels of activity.

Although no single factor or condition can be said to determine when an animal will go into hibernation, specific changes include an increase in the quantity of magnesium in the blood and a reduction in the activity of endocrine glands, such as the pituitary, thyroid, and adrenals. A reduction in gonadal activity has also been observed. Hibernation does not occur when the gonads are in an actively functional state. Perpetuation of the species requires that the animal be warm and active during the mating and pregnancy periods.

There appears to be a relationship between sleep and hibernation with available evidence suggesting that hibernation is entered into from a state of sleep. To consider hibernation a form of sleep, it must be considered a remarkably complex one. Hibernation and sleep are somewhat similar in that essential body processes continue during both periods at a lowered level. In sleep, the heart beats less rapidly; breathing is slower; and the body produces less heat, necessitating that a sleeping person be protected from the cold.

Hibernation in Birds

Birds normally have higher and more variable temperatures than do mammals. Whereas mammalian temperatures normally range between 36 and 39 °C (97 and 102 °F), avian temperatures range between 37.7 and 43.5 °C (99.9 and 110.3 °F), with the majority between 40 and 42 °C (104 and 108 °F). Although the nesting temperature of most passerine species (perching songbirds) is about 40.5 °C (104.9 °F), primitive bird species—like primitive mammals—have lower temperatures than do the more advanced species. The kiwi, for example, has an average coenothermic body temperature of 37.8 °C (100 °F). In general, the temperatures of small birds fluctuate more than do those of large birds. The temperature of a house wren (*Troglodytes*) may fluctuate 8 °C (14 °F) in 24 hours, that of a robin (*Turdus*) fluctuates about 6 °C (11 °F), and that of the domestic duck only about 1 °C (2 °F).

The circadian period of activity and rest in birds is accompanied by a temperature cycle. Birds active in the daytime have their highest temperatures late in the afternoon and their lowest in the early morning. Nocturnal species, however, such as owls and the kiwi, have their maximum body temperatures at night, when they are most active. Seasonal temperature variations are also found in birds, and, like mammals, certain birds exhibit thermic instability. Although some are capable of maintaining a highly stable body temperature, others have a fluctuating body temperature. A torpid poorwill (*Phalaenoptilus nuttallii*) is an example of a bird that demonstrates both thermic instability and true hibernation. Although its coenothermic body temperature is relatively constant, it can, through the influence of a thermoregulatory centre (the hypothalamus) in the floor of the brain, become essentially poikilothermous. Under such influence, its body temperatures approximate those of the environment.

Considering that hibernation and estivation are devices to avoid such factors as stressful extremes of temperature, lack of water, unavailability of food, or lessened photoperiod, they also must be energy-conservation devices for the animals concerned. Even short periods of torpor can conserve energy. The efficiency of this energy-conservation system can be demonstrated by comparing the smallest bird, the hummingbird, which exhibits circadian torpor, with the shrew, the smallest mammal, which remains active throughout a 24-hour period. Oxygen consumption is an indicator of metabolic rate, and at an environmental temperature of 24 °C (75 °F) during the day, an awake but resting hummingbird consumes about 14 millilitres of oxygen per gram per hour. At dusk, the rate drops first to a sleeping level and then plunges to a torpid level of about 0.8 millilitre of oxygen per gram per hour. Just before daybreak, the bird awakens for another activity period. The hummingbird has the highest metabolic rate and the greatest metabolic range of any vertebrate. The shrew, in contrast, consumes about the same amount of oxygen as the hummingbird does during the day and even increases the amount slightly at night.

The hummingbird uses about 10.3 calories (units of heat energy) during each 24-hour period if it sleeps at night without becoming torpid but only 7.6 calories if it becomes torpid. As it wakes from the torpid state, its temperature increases about 1 °C (2 °F) per minute to a maximum; the entire process takes less than 30 minutes and sometimes as little as 10 minutes. The energy required to warm the tissues of the hummingbird is relatively small; a hummingbird that weighs four grams (.141 ounces) expends only 0.114 calorie to warm its body from 10 to 40 °C (50 to 104 °F). This is only $^{1}/_{85}$ of the total 24-hour expenditure of energy of a hummingbird in nature.

The behaviour of the hummingbird can be contrasted to that of a larger bird, such as the poorwill, which is a nocturnal, insect-catching bird. During an average 24-hour day, the poorwill has brief periods of activity at dusk and just before dawn, the total of which is scarcely more than an hour. The temperature of the poorwill during these periods of activity, which are correlated with the bird's feeding habits, is between 40.5 and 43.1 °C (104.9 and 109.6 °F). Between periods of activity, the bird rests quietly, and its body temperature drops 1 to 3 °C (2 to 5 °F).

During periods when a supply of flying insects is not available, the bird hibernates in depressions in rocks or other suitably protected places, to which it returns each year. When hibernating, the bird's temperature is frequently within 1 °C (2 °F) of that of the environment, thus saving a great deal of energy. A poorwill whose body temperature is 5 °C (41 °F) has a metabolic rate only 3 percent of its metabolic rate at normal body temperature. Because the poorwill is a larger bird than a hummingbird, it may take more than an hour for it to emerge from hibernation.

Hibernation in Mammals

It takes longer for larger animals than for smaller ones to go into hibernation because heat must radiate from the body before the temperature can be lowered. Thus, it would require a considerable amount of time for large birds or mammals to go into and emerge from hibernation each day, as do bats and hummingbirds. A 200-kilogram (440-pound) bear, for example, would need 5,100 calories to warm from 10 to 37 °C (50 to 99 °F). Unlike the hummingbird, which uses only 1/85 of its total daily energy expenditure to emerge from hibernation, the amount expended by a bear would be equivalent to its full 24-hour energy budget. Even if there were enough time in 24 hours for a large animal to enter into and emerge from dormancy, therefore, it would be metabolically extravagant, thus defeating a purpose of hibernation.

Actually, the most common misapplication of the term hibernation is in relation to the bear, which is not a true hibernator. Its body temperature, which normally averages 38 °C (100 °F), drops during its winter lethargy to about 34 °C (93 °F), seldom getting below 31.2 °C (88.2 °F). Hence, a bear's temperature during the winter does not approximate that of the environment. This is indicative of winter rest rather than true hibernation. During this inactive period, the bear sleeps but is, nonetheless, warm and capable of activity when stimulated, unlike a true hibernator. Moreover, it is also during this period when females give birth to cubs that suckle and are maintained by maternal warmth until they emerge from the den in the spring. This behaviour contrasts with that of the Arctic ground squirrel, whose normal temperature is the same as that of the bear but whose temperature during hibernation drops to near freezing and, in some cases, to a degree or two below 0 °C (32 °F).

Although certain mammals are said to hibernate, they do not necessarily enter a state of deep hibernation during winter. Instead, for weeks at a time they may be inactive and lethargic in behaviour, with a slightly depressed body temperature. The chipmunk (*Eutamias*) is an example of what has been termed a shallow hibernator, as are bears and raccoons. Superficial hibernation, apparently a compromise between the minimum energy requirements of a deep hibernator and the high energy expended by an animal that remains active during the winter, saves energy without the stress of hibernation. The animal can thus conserve food, while still being able to escape from predators and such dangers as flooding of its burrow. The main energy source during the winter in this shallow hibernation state is food stored in the winter nest. There are instances, however, of shallow hibernators, such as the chipmunk, that enter a state of deep hibernation, particularly if without food.

Omitting the thermally unstable mammals, the true mammalian hibernators are those whose lowered body temperatures approximate that of the environment and those who require extensive and complex physiological changes to turn from a warm-blooded animal into an essentially cold-blooded one for an appreciable length of time. Only three orders of placental mammals display such behaviour: the Insectivora, as exemplified by the hedgehog; the Chiroptera, the bats; and the Rodentia, including the marmot, hamster, dormouse, hazel mouse, and ground squirrel.

A typical mammalian hibernator, such as the Arctic ground squirrel, finds a protected environmental niche—in this case, a burrow beneath the surface—and builds a nest of grass, hair, and other materials to provide still further insulation. The usual hibernating position is one of being curled up in a ball with the extremities tucked tightly against the body so there is a minimal surface-to-volume ratio. After the temperature of the animal has dropped near that of the ambient temperature, it appears to be dead: its respiration is imperceptible, about three irregular breaths per minute; it does not react to outside stimuli in an observable fashion; nor does it react to being handled and uncurled, although such handling will trigger wakening mechanisms.

The internal organs, such as the digestive tract and the endocrine glands, are almost totally inactive. Because the process of hibernation necessitates the mobilization of all body resources, it places great demands on the tissues, all of which are directed toward the problem of maintaining the animal's metabolism at the minimal level necessary for life during the hibernating period. This means that all activity not immediately germane to the process of living at the lowest possible metabolic level ceases. Even bones and teeth deteriorate during hibernation. The hibernator apparently is balanced on a very narrow line between the maintenance of life at a level that makes recovery from hibernation possible and a reduction of metabolism to a level that will lead to death. Evidence obtained from tissues indicates that the process of hibernation is a precarious method of survival at best and one from which many animals do not awaken. As a mechanism of species survival, hibernation seems effective; for the survival of the individual, however, it is an uncertain and dangerous process.

The hibernator does not remain in a continuous state of hibernation from the time it enters in the fall until it emerges in the spring. Hibernating Arctic ground squirrels, for example, awaken at intervals of every three weeks or less. During this time the animals may move about and sometimes emerge from the burrow. These periods of arousal are more frequent at the beginning and end of a hibernation period than in mid-hibernation. The lower the temperature at which an animal hibernates, the fewer the awakenings.

During the period of hibernation about 40 percent of the total body weight is lost, an average of about 0.2–0.3 percent per day. One period of arousal and wakefulness consumes more heat and energy than many days in hibernation. About 90 percent of the total heat production and weight loss during hibernation takes place during the arousal periods; only 10 percent is required to maintain the animal in hibernation. Thus, in the case of an unusually long or hard winter, the animal may be called upon to use all of its available energy sources in periodic arousals; it then enters one final hibernation period from which it does not awaken. Animals that store food in the nest have a chance to renew their energy requirements by eating when they awaken periodically.

Hibernating mammals can be divided into four major groups according to the way they enter hibernation. One group is exemplified by the golden hamster, which waits a variable time of from one to three months in the cold and then enters hibernation in one major temperature reduction. This is accomplished when the biochemical and physiological preparations have been sufficient to lower the animal to a level at which it is receptive to the hibernating stimulus, which causes the abandonment of the temperature differential between ambient and body temperatures.

A second group, of which the pocket mouse (*Perognathus*) is an example, prepares for hibernation relatively rapidly, waiting only a few days before becoming torpid in one major temperature decline. The third group, which constitutes most of the mammalian hibernators, includes ground squirrels and marmots. These animals wait only a few days before entering hibernation and then go through a series of steps of torpor and arousal, each one at successively lower body temperatures, until the level dictated by the stage of preparation for hibernation is reached.

The fourth group, which includes most of the bats, becomes inactive in the poikilothermous manner; that is, the body temperature follows the ambient temperature. Even though the bat seems ready to hibernate at any season, survival during hibernation depends upon more adequate preparation than is necessary for the transitory periods of torpor. Bats not only exhibit true hibernation during the winter but also have natural periods of hypothermia (subnormal temperature), which are unrelated to hibernation, during the remainder of the year.

The woodchuck, the dormouse, and the California ground squirrel enter hibernation in successive stages, with a complete or nearly complete awakening between each one. In the woodchuck, an initial decline in temperature is followed by an arousal. During the second decline there is a lower and more pronounced fall in body temperature, followed by a less pronounced rise. This process continues until the body temperature is essentially the same as that of the environment.

Physiological Changes During Mammalian Hibernation

The body temperature of a hibernating mammal is affected by changes in respiration, heart rate, and oxygen consumption; all are apparently mediated by a part of the nervous system. The heart rate decreases prior to a decline in body temperature. In the woodchuck, the rate may drop from 153 to 68 heartbeats per minute within 30 minutes. In the California ground squirrel, the heart may beat as slowly as once a minute at 5 °C (41 °F). In contrast, the hearts of non-hibernators generally will not beat at all at temperatures below 10–20 °C (50–70 °F).

As an Arctic ground squirrel prepares for hibernation, its heart rate and its blood pressure decrease. Both may be detected before a decrease in body temperature can be noted. When the animal enters hibernation, temperatures of both the heart and abdominal regions are identical, indicating an even blood flow between the anterior (front) and posterior (rear) parts of the body. As the body temperature drops, the resistance to blood flow in the peripheral parts of the circulatory system increases because of the increased viscosity (resistance to flow) of the chilled blood and the constriction of the distal arterioles (small arteries) of the body. This peripheral resistance maintains blood pressure at relatively high levels in the deeply hibernating squirrel, even when the heart beats only three or four times a minute.

The nervous system of hibernators also is acclimated, certain specific structures and pathways seemingly maintained to regulate and coordinate metabolism as temperatures drop. This adaptation of the nervous system enables changes in the environment to be perceived, even when the animal is torpid. In the Arctic ground squirrel, measurements of the general electrical activity of the brain indicate a 90 percent reduction when the animal is in hibernation, at which time brain temperatures approximate 6 °C (43 °F). During hibernation, both the peripheral nervous system (all the nerves outside the brain and spinal cord, which constitute the central nervous system) and the spinal cord have an increased sensitivity to certain stimuli; in addition, the areas of the brain that regulate temperature as well as cardiac (heart) and respiratory function remain active at ambient temperatures, below which the mammalian nervous system normally ceases to function.

Changes in the circulatory system involving constriction (narrowing) of posterior vessels and the favouring of anterior circulation allow the brain temperature of hibernators to remain a few degrees warmer than the environmental level. This enables the temperature of the brain to remain constant despite fluctuations in the temperature of the skin.

The male sex hormone testosterone stimulates reproductive activity. The golden hamster will not hibernate if injected with more than five milligrams of a hormonal preparation. Hibernation is also prevented if the animal is fed or injected with thyroid hormones or thyroid-stimulating extracts. The latter would seem to implicate the thyroid as another endocrine gland that plays an important role in hibernation. There is, in fact, a seasonal progression and regression of thyroid activity in hibernators; maximal activity occurs in the spring and minimal activity in the fall. And because hibernation does not take place in the absence of the adrenal glands, it appears that a minimal adrenal activity is also necessary for hibernation and survival.

The importance of timing in the annual rhythm of activity and dormancy can be demonstrated: when hibernators are exposed to cold temperatures in spring and summer, they react as do all homoiotherms by increasing their thyroid activity and metabolic rate to maintain normal body temperature. But if they are exposed to cold temperatures in the fall, the thyroid activity and metabolic rate of hibernators are lowered. In some species, a combination of decreased food and lower ambient temperature is required to reduce activity of the thyroid gland and to produce hibernation, although cold alone is sufficient in ground squirrels and the dormouse.

Although hibernation does not take place during periods of gonadal activity or stimulated thyroid activity, it can occur during increased activity of the pituitary gland. This would suggest that there is a dissociation of cellular growth and hormone synthesis that is normally controlled by hormone secretion of the pituitary and its target organs. Thus, the triggering mechanism for the resumption of normal endocrine activity apparently resides elsewhere than in the pituitary. The function of the hypothalamic region of the brain in regulating appetite, fat deposition, water intake, and diuresis (increased excretion of urine), as well as in the control of temperature and sleep, would appear to make it a key area in directing life processes of the hibernator. Furthermore, the fact that the hypothalamus regulates the pituitary and other endocrine glands not only supports this thesis but also indicates that this area of the brain is the prime, or master, regulator of the entire hibernation process.

The Arctic ground squirrel may spend more than half its life in hibernation. It thus must be able to breed, rear young, maintain its home burrow, and prepare for the period of hibernation during an activity period of less than six months. This requires considerable adaptation of both metabolic and behavioral patterns. Prior to entering hibernation in late September or early October, there is a renewal of sexual activity in the testes of males, and, throughout the period of hibernation, they continue to grow.

MARMOT

Marmots (*Marmota*) are any of 14 species of giant ground squirrels found primarily in North America and Eurasia. These rodents are large and heavy, weighing 3 to 7 kg (6.6 to 15.4 pounds), depending on the species. Marmots are well suited for life in cold environments and have small fur-covered ears, short, stocky legs, and strong claws for digging. Length of the bulky body is 30 to 60 cm (11.8 to 23.6 inches), and the short, bushy tail is 10 to 25 cm (3.9 to 9.8 inches) long. Their long, thick fur is slightly coarse and may be yellowish brown (usually frosted with buff white), brown, reddish brown, black, or a mixture of gray and white.

Marmots are found north of Mexico and in Eurasia from the European Alps through north-central Asia, the Himalayas, and northeastern Siberia to the Kamchatka Peninsula. They inhabit open country in mountains and plains, preferring montane meadows, steppes, tundra, and forest edges. All live in burrows that they excavate, and most mountain species construct burrows beneath boulder fields, rocky slopes, and crevices in cliff faces. This terrain provides protection from predators such as grizzly bears, which are aggressive diggers and a significant predator of the Alaska marmot (*Marmota broweri*) in the Brooks Range. Rocks and cliffs also serve as observation sites where the rodents sit upright watching for both terrestrial

Although marmots hibernate deeply, they may wake and leave their burrows on warmer days. E.R. Degginger

and aerial predators. When alarmed, marmots emit a sharp, piercing whistle and scurry to their burrows if danger persists.

Marmots are active during the day (diurnal) and are almost entirely vegetarian. The Alaska marmot, which grazes on low-nutrient tundra vegetation, must seek productive foraging areas where it competes indirectly with other mammalian grazers, including caribou, Dall's sheep, and voles. Some marmots, such as the Alpine marmot (*M. marmota*) and the hoary marmot (*M. caligata*) of northwestern North America, are gregarious and social, but others, including the woodchuck (*M. monax*) of Canada and the United States, are solitary. All hibernate in winter, most of them deeply, although some may emerge from their burrows for short periods on mild winter days. During hibernation they live on fat reserves accumulated during the summer. The hoary marmot hibernates for up to nine months, its fat reserves amounting to 20 percent of its total body weight. Marmots mate soon after they emerge from hibernation. Gestation lasts about a month, and a litter of generally 4 or 5 (recorded extremes range from 2 to 11) is born in a nest within the burrow. Most marmots produce young every year, but the Olympic marmot (*M. olympus*) of the Olympic Mountains in the United States bears young every other year.

Marmots belong to the squirrel family (Sciuridae) within the order Rodentia. The closest living relatives of marmots are ground squirrels and prairie dogs. Marmots' evolutionary history is recorded in North America by fossils of extinct species from the late Miocene Epoch (13.8 million to 5.3 million years ago). In Eurasia there is no evidence earlier than the Pleistocene Epoch (2.6 million to 11,700 years ago).

On the Arctic slope in early May, the male ground squirrel emerges from its burrow. As it uses the remaining fat and eats the stores of seeds and other food still in the nest, the male reaches a period of reproductive readiness. Mating takes place in the middle of May, and the young are born in the middle of June, after a gestation period of about 25 days. By the middle of July the young are above ground and eating the green Arctic vegetation, which they continue to eat until the onset of hibernation. By October, both the young of the year and the adults from the previous year weigh nearly 1,000 grams (2.2 pounds).

In the bat, the reproductive cycle is interrupted by hibernation. Gonadal activity in the male reaches its maximum in the fall, when copulation with the female occurs. The animals then hibernate, and the production of sperm in the male ceases. Sperm deposited in the female are stored in her reproductive tract throughout the period of hibernation. Fertilization occurs the next spring, when the eggs are ovulated (released from the ovaries) within a few days after awakening from hibernation.

The only exception to the general hibernation–reproduction pattern of bats is the vespertilionid bat (*Miniopterus*), in which there is no delayed ovulation and fertilization. In this species the eggs are ovulated soon after copulation, in the fall, and fertilized immediately. During the ensuing period of hibernation embryonic development is initiated and slowed, but it does not actually cease. The young are born in the early summer, soon after hibernation ends. The introduction of hibernation during pregnancy makes the gestation period several months longer than in non-hibernating tropical members of the same genus.

Cyclical reproductive activity has thus become adapted to the shortened activity season available to the hibernator. But although the annual sequence of reproductive events is known, the external stimuli that regulate the reproductive cycles of bats and other hibernators are unknown. More knowledge is needed concerning the endocrine and nervous mechanisms that presumably regulate reproductive processes internally. It has been suggested that the pituitary–gonadal relationship influences the hibernating cycles as well as the reproductive cycle, hence both the latter and homoiothermism are controlled by a common mechanism. Such a suggestion is attractive in that the mechanism solves the regulation problems, but more needs to be known of the way in which hibernation directly or indirectly modifies the action of endocrine and neural mechanisms that direct the reproductive cycle.

Hibernating organisms have a certain degree of resistance to infectious diseases that appears to be attributable to at least three factors, all of which are related to temperature. One is the fact that the lowered temperature of the host and the commensurate slowing of its metabolic processes prevent the multiplication of parasites to a greater extent than they affect the host's defensive mechanisms. Second, lower temperatures are more harmful to the development of a disease organism than to the host, as has been shown with the parasite *Trichinella spiralis*. In bats hibernating at 5 °C (41 °F), only larvae have been recovered from the intestines; but mature adult worms have been recovered from the intestines of bats kept at 35 °C (95 °F). The third factor is that the influence of low temperature on the chemical composition of the host tissues may also affect infectious organisms.

Hibernation also seems to protect animals from radiation. When ground squirrels are irradiated with radioactive cobalt while hibernating, they are found to be more resistant to the effects of the radiation than are squirrels irradiated while warm and active. This resistance, which is apparent over a wide range of doses, suggests that protective mechanisms function in the hibernating animal. In both hibernating and non-hibernating animals, repair processes within cells occur the first day after irradiation; however, when the metabolic requirements of cells are small, as in hibernation, the injured cells seem to be more capable of repair, and survival is greater. The large metabolic requirements imposed on injured cells of warm and active animals appear to render them incapable of an adequate repair response.

The process of awakening in the Arctic ground squirrel takes about three hours. There is a rapid rise in heartbeat and a decrease in peripheral circulatory resistance; the area around the head and heart warms more rapidly than the posterior part of the animal. This differential vasodilatation (widening of the blood vessels) in the anterior part of the body is a unique and vital part of the awakening process. The concentration of active circulation in this region results in a high blood pressure and an efficient and rapid warming. If a drug is administered during awakening that causes vasodilatation throughout the body, there is a marked drop in blood pressure even though the heart may almost double its rate; thus, the heart cannot maintain a high blood pressure at this time if all blood vessels are dilated. Later during the arousal process, after the anterior part of the body has been warmed, the posterior part of the animal warms rapidly.

Despite the deterioration of glands and tissues and the drastic reduction of all metabolic activity during hibernation, within 24 hours after arousal, all the squirrel's physiological processes are essentially normal. This rapid repair and recovery mechanism is one that requires further study.

CHAPTER 7
NOTABLE FIGURES IN CONSERVATION

Many researchers, government officials, and activists have made significant contributions to the study of conservation and ecology. Some of the more notable contributors are described in this chapter.

RACHEL CARSON

(b. May 27, 1907, Springdale, Pa., U.S.—d. April 14, 1964, Silver Spring, Md.)

Rachel Carson was an American biologist well known for her writings on environmental pollution and the natural history of the sea.

Carson early developed a deep interest in the natural world. She entered Pennsylvania College for Women with the intention of becoming a writer but soon changed her major field of study from English to biology. After taking her bachelor's degree in 1929, she did graduate work at Johns Hopkins University (M.A., 1932) and in 1931 joined the faculty of the University of Maryland, where she taught for five years. From 1929 to 1936 she also taught in the Johns Hopkins summer school and pursued postgraduate studies at the Marine Biological Laboratory in Woods Hole, Mass.

In 1936 Carson took a position as aquatic biologist with the U.S. Bureau of Fisheries (from 1940 the U.S. Fish and Wildlife Service), where she remained until 1952, the last three years as editor in chief of the service's publications. An article in *The Atlantic Monthly* in 1937 served as the basis for her first book, *Under the Sea-Wind*, published in 1941. It was widely praised, as were all her books, for its remarkable combination of scientific accuracy and thoroughness with an elegant and lyrical prose style. *The*

Bob Hines and Rachel Carson conducting research in Florida, 1952. U.S. Fish and Wildlife Service

Sea Around Us (1951) became a national best seller, won a National Book Award, and was eventually translated into 30 languages. Her third book, *The Edge of the Sea*, was published in 1955.

Carson's prophetic *Silent Spring* (1962) was first serialized in *The New Yorker* and then became a best seller, creating worldwide awareness of the dangers of environmental pollution. The outlook of the environmental movement of the 1960s and early '70s was generally pessimistic, reflecting a pervasive sense of "civilization malaise" and a conviction that the Earth's long-term prospects were bleak. *Silent Spring* suggested that the planetary ecosystem was reaching the limits of what it could sustain. Carson stood behind her warnings of the consequences of indiscriminate pesticide use, despite the threat of lawsuits from the chemical industry and accusations that she engaged in "emotionalism" and "gross distortion." Some critics even claimed that she was a communist. Carson died before she could see any substantive results from her work on this issue, but she left behind some of the most influential environmental writing ever published.

BARRY COMMONER

(b. May 28, 1917, Brooklyn, N.Y., U.S.)

Barry Commoner is an American biologist and educator. He studied at Harvard University and taught at Washington University and Queens College. His warnings, since the 1950s, of the environmental threats posed by modern technology (including nuclear weapons, use of pesticides and other toxic chemicals, and ineffective waste management) in such works as his classic *Science and Survival* (1966) made him one of the foremost environmentalist spokesmen of his time. He was a third-party candidate for U.S. president in 1980.

PAUL R. EHRLICH

(b. May 29, 1932, Philadelphia, Pa., U.S.)

Paul R. Ehrlich is an American biologist and educator who in 1990 shared Sweden's Crafoord Prize (established in 1980 and awarded by the Royal Swedish Academy of Sciences to support those areas of science not covered by the Nobel Prizes) with biologist E.O. Wilson.

Ehrlich received early inspiration to study ecology when in his high school years he read William Vogt's *Road to Survival* (1948), an early study of the problem of rapid population growth and food production. Ehrlich graduated in zoology from the University of Pennsylvania (B.A., 1953) and took M.A. and Ph.D. degrees from the University of Kansas (1955, 1957). He held a few research positions before accepting (1959) a position at Stanford University, where he became a full professor of biology from 1966 and Bing professor of population studies from 1976. Though much of his research was done in the field of entomology, Ehrlich's overriding concern became unchecked population growth. He was concerned that humanity treat the Earth as a spaceship with limited resources and a heavily burdened life-support system; otherwise, he feared, "mankind will breed itself into oblivion." He published a distillation of his many articles and lectures on the subject in *The Population Bomb* (1968) and wrote hundreds of papers and articles on the subject.

AL GORE

(b. March 31, 1948, Washington, D.C., U.S.)

Albert Arnold Gore, Jr., was the 45th vice president of the United States (1993–2001) in the Democratic administration of President Bill Clinton. In the 2000 presidential election, one of the most controversial elections in American history, Gore won the nationwide popular vote over George W. Bush by more than 500,000 votes but narrowly lost in the electoral college, 271–266—the first inversion of the electoral and popular vote since 1888. In 2007 Gore was awarded, with the United Nations' Intergovernmental Panel on Climate Change, the Nobel Prize for Peace for his efforts to raise awareness about global warming.

The son of a Democratic congressman and senator from Tennessee, Gore graduated from Harvard University in 1969 and enlisted in the army, serving in the Vietnam War as a military reporter from 1969 through 1971. He then became a reporter for *The Tennessean*, a newspaper based in Nashville, Tennessee. While working (1971–76)

Al Gore, 1994. U.S. Department of Defense

for that paper, Gore also studied philosophy and law at Vanderbilt University.

Gore won election to the U.S. House of Representatives in 1976 and was reelected three times before winning a seat in the Senate in 1984. In 1988 he was an unsuccessful candidate for the Democratic presidential nomination. Gore was reelected to the Senate in 1990, and in 1991 he was one of only 10 Democratic senators who voted to authorize the use of American military force against Iraq in the Persian Gulf War. In 1992 he was chosen by Clinton, the Democratic presidential nominee, to be his running mate, and Gore became vice president when Clinton defeated Republican incumbent George Bush in the 1992 presidential election. In 1993 Gore helped the Clinton administration secure congressional passage of the North American Free Trade Agreement. Gore and Clinton were reelected in 1996 to a second term, defeating the Republicans led by Bob Dole.

Gore announced his candidacy for the presidency of the United States in June 1999. A moderate Democrat, his campaign focused on the economy, health care, and education. On issues that were controversial in the United States, Gore generally supported the Democratic Party's platform, favouring abortion rights for women and greater restrictions on guns, but he broke with the party's traditional stance on the death penalty, which he supported. Gore favoured strong measures to protect the environment, setting out his ideas on this issue in his book *Earth in the Balance: Ecology and the Human Spirit* (1992). Campaign finance reform also became a prominent issue following allegations that Gore conducted illegal fund-raising activities during Clinton's 1996 reelection bid. Although Gore denied knowing at the time that he was committing any wrongdoing, Republicans repeatedly called for an independent counsel to investigate the matter.

After easily winning the Democratic presidential nomination, Gore trailed in most public opinion polls until his selection of a vice presidential running mate, Senator Joseph Lieberman, who became the first Jewish American on a national presidential ticket. As the campaign progressed, the polls remained close. When the votes were counted on election night, it became clear that the election would turn on the outcome in Florida, whose 25 electoral votes would give the winner of that state a narrow majority in the electoral college. After the television networks declared Bush the winner in Florida and president-elect, Gore called Bush to concede. Later that evening, however, as continuing returns from Florida showed Gore closing the gap with Bush, Gore decided to cancel a public concession speech and called Bush to withdraw his concession. Trailing in Florida after a mandatory recount by fewer than 1,000 votes, Gore sought a manual recount of ballots in heavily Democratic counties in southern Florida. For five weeks, the election remained unresolved as Florida state courts and federal courts weighed arguments by the Bush and Gore campaigns. Eventually, Gore won a controversial 4–3 victory in the Florida Supreme Court, which ordered a statewide manual recount of the approximately 45,000 "undervotes" (i.e., ballots that machines recorded as not clearly expressing a presidential vote). The Bush campaign quickly filed an appeal with the U.S. Supreme Court. Concluding (7–2) that a quick statewide recount could not be performed fairly unless elaborate ground rules were established, the court issued a controversial 5–4 decision to reverse the Florida Supreme Court's recount order, effectively awarding the presidency to Bush. In a speech watched by more than 65 million Americans the following evening, Gore conceded defeat, saying that "while I strongly disagree with the court's decision, I accept it." He also pledged to "honour the new president-elect and do everything possible to help him bring Americans together in fulfillment of the great vision that our Declaration of Independence defines and that our Constitution affirms and defends."

Gore subsequently devoted much of his time to environmental issues. He discussed global warming in the 2006 documentary *An Inconvenient Truth* and in its companion book. The film won an Academy Award for best documentary. In 2007 Gore published *The Assault on Reason*, in which he sharply criticized the administration of President Bush. Later that year he received an Emmy Award for creative achievement in interactive television for Current TV, a user-generated-content channel he cofounded in 2005.

THEODORE JESSE HOOVER

(b. Jan. 28, 1871, West Branch, Iowa, U.S.—d. Feb. 4, 1955, near Santa Cruz, Calif.)

Theodore Jesse Hoover was an American mining engineer, naturalist, educator, and elder brother of U.S. Pres. Herbert Hoover.

Hoover was the oldest of three children born to Jesse Clark Hoover, a village blacksmith and dealer in agricultural machinery, and Huldah Randall Minthorn Hoover, a teacher and Quaker minister. He was educated at Friends Pacific Academy, a Quaker school in Newberg, Ore., and at Penn College in Iowa. He received his A.B. degree in geology and mining from Stanford University in 1901. From 1903 to 1919 he was manager or consulting engineer in the gold mines of California, western Australia, Mexico, and Alaska; was an engineer or administrator of lead and silver mines in Burma and of copper mines in Finland and Russia; and had offices in London and San Francisco.

In 1919 he returned to Stanford University as a professor of mining and metallurgy, serving as dean of the school of engineering from 1925 until his retirement in 1936. Thereafter he lived at his ranch (the Rancho del Oso) in Santa Cruz county, Calif. As a writer, conservationist, and naturalist, he directed a definitive study of the life cycle of the Pacific Coast steelhead trout, assisted by the California fish and game commission. He was the author of *Concentrating Ores by Flotation* (1912), *Economics of Mining* (1933), and *The Engineering Profession* (1941) and of numerous articles in technical publications. He was a member of the American Society of Mining and Metallurgical Engineers and a founding member of the Cooper Ornithological Society.

In 1905 Hoover Lake in Santa Clara county, Calif., was named after him. He explored and mapped the area around the lake during the summers of 1904 and 1905 while serving as manager of the Standard Consolidated Mines. Also named in his honour is the Theodore J. Hoover National Preserve in northern Santa Cruz county. The preserve is noted for containing one of the rarest coastal marsh habitats in central California and sheltering a variety of rare and endangered species.

ALDO LEOPOLD

(b. Jan. 11, 1887, Burlington, Iowa, U.S.—d. April 21, 1948, near Madison, Wis.)

Aldo Leopold was an American environmentalist. After attending Yale University, he worked for the U.S. Forest Service (1909–28), mainly in the Southwest. In 1924 the country's first national wilderness area (Gila Wilderness Area in New Mexico) was created at Leopold's urging. From 1933 to 1948 he taught at the University of Wisconsin. A fervent campaigner for the preservation of wildlife and wilderness areas, he was a director of the Audubon Society from 1935 and became a founder of the Wilderness Society in the same year. His *Game Management* (1933) was followed in 1949 by the posthumous *A Sand County Almanac*, which eloquently called for the preservation of ecosystems. Read by millions, it strongly influenced the budding environmental movement.

JANE LUBCHENCO

(b. Dec. 4, 1947, Denver, Colo., U.S.)

Jane Lubchenco is an American environmental scientist and marine ecologist who became the first woman to serve as administrator of the National Ocean and Atmospheric Administration (NOAA) and as undersecretary for oceans and atmosphere (2009–).

Lubchenco grew up in Denver and received a bachelor's degree in biology (1969) from Colorado College. She obtained a master's degree in zoology (1971) from the University of Washington and a doctorate in ecology (1975) from Harvard University. Her thesis work focused on community structure in coastal rockpools. She served as an assistant professor at Harvard from 1975 to 1977. In 1977 she began teaching marine biology at Oregon State University and the following year became a research associate at the Smithsonian Institution, a position she held until 1984. Her areas of research included algal ecology, plant-herbivore and predator-prey interactions, global change community structure, and the evolutionary ecology of individuals. She continued to teach at Oregon State, receiving an endowed chair in 1995.

Lubchenco served as president of the Ecological Society of America (1992–93). Later, she was chair of the task force on the environment (1998–2000) at the

National Science Board and an adviser (1996–2000) to Religion, Science, and the Environment, a cross-disciplinary partnership of scientists and religious leaders. She was president of the International Council for Science from 2002 to 2005. Lubchenco held memberships in the National Academy of Sciences (1996), the American Philosophical Society (1998), and other prestigious organizations.

Recognizing that environmental change did not come about without mass participation, Lubchenco sought ways to better inform the public of scientific issues and to bridge the gulf between researchers and the rest of the world. In a 1997 speech she proposed the idea of a social contract between scientists and society. Lubchenco founded the Aldo Leopold Leadership Program, aimed at enhancing the ability of research scientists to communicate their findings to a general audience, in 1998, and she helped to create the Communication Partnership for Science and Sea (COMPASS), an organization devoted to educating policy makers on ocean ecology, in 1999. That year she also helmed the Partnership for Interdisciplinary Studies of Coastal Oceans (PISCO). Lubchenco was one of the primary organizers (2008–09) of Climate Central, which focused on disseminating information on climate change to the public.

Her ability to combine passionate advocacy with pragmatism led to Lubchenco's nomination as NOAA administrator and undersecretary of commerce for oceans and atmosphere by Pres. Barack Obama in December 2008. Though some fishing industry officials expressed reservations at her nomination because of the attention she had called to the problem of fishery collapse, Lubchenco was confirmed in March 2009 to wide approval from the scientific community.

JOHN MUIR

(b. April 21, 1838, Dunbar, East Lothian, Scot.—d. Dec. 24, 1914, Los Angeles, Calif., U.S.)

John Muir was an American naturalist and advocate of U.S. forest conservation. He was largely responsible for the establishment of Sequoia and Yosemite national parks in California.

Muir emigrated from Scotland with his family to a farm near Portage, Wis., in 1849. After attending the University of Wisconsin, Madison, he worked on mechanical inventions, but in 1867, when an industrial accident nearly cost him an eye, he abandoned that career and devoted himself to nature. He walked from the Middle West to the Gulf of Mexico, keeping a journal, *A Thousand-Mile Walk to the Gulf* (published posthumously in 1916) as he went. In 1868

Yosemite National Park, California. © Digital Vision/Getty Images

he went to the Yosemite Valley, California, from which he took many trips into Nevada, Utah, Oregon, Washington, and Alaska, inspired by his interest in glaciers and forests. He was the first to attribute the spectacular Yosemite formations to glacial erosion, a theory now generally accepted.

As early as 1876, Muir urged the federal government to adopt a forest conservation policy. The Sequoia and Yosemite national parks were established in 1890. Early in 1897, Pres. Grover Cleveland designated 13 national forests to be preserved from commercial exploitation. Business interests induced Congress to postpone the effect of that measure, but two eloquent magazine articles (June and August 1897) by Muir swung public and Congressional opinion in favour of national forest reservations. He also influenced the large-scale conservation program that was initiated by Pres. Theodore Roosevelt, who in 1903 accompanied Muir on a camping trip to the Yosemite region. In 1908 the government established the Muir Woods National Monument in Marin County, California.

GIFFORD PINCHOT

(b. Aug. 11, 1865, Simsbury, Conn., U.S.—d. Oct. 4, 1946, New York, N.Y.)

Gifford Pinchot was a pioneer of U.S. forestry and conservation.

Pinchot graduated from Yale in 1889 and studied at the National Forestry School in Nancy, France, and in Switzerland, Germany, and Austria. Upon his return home in 1892, he began the first systematic forestry work in the United States at Biltmore, the estate of George W. Vanderbilt, in North Carolina. In 1896 he was made a member of the National Forest Commission of the National Academy of Sciences, which worked out the plan of U.S. forest reserves, and in 1897 he became confidential forest agent to the Secretary of the Interior. In 1898 he was appointed chief of the Division, later Bureau, of Forestry and then the Forest Service (created 1905) in the Department of Agriculture, which office he held under Presidents William McKinley, Theodore Roosevelt, and William Howard Taft, until 1910. During his administration the entire forest-service system and administrative machinery were built up, and Pinchot's enthusiasm and promotional work did much for the conservation movement in general. He also served as a member of the Public Lands Commission, which he initiated in 1903, and the Inland Waterways Commission (1908). In 1908 he became chairman of the National Conservation Commission. He founded the Yale School of Forestry at New Haven, Conn., as well as the Yale Summer School of Forestry at Milford, Pa., and in 1903 became professor of forestry at Yale. In 1920 he was appointed state forester of Pennsylvania and began a systematic administration of the forest areas of that state.

With Theodore Roosevelt, Pinchot helped to found the Bull Moose Party in 1912. From 1923 to 1927 and from 1931 to 1935 he was governor of Pennsylvania. In his first term he forced a reorganization of the state government and the establishment of a budget system. He settled a coal strike by arbitration in 1923. Pinchot's autobiography, *Breaking New Ground,* was published posthumously in 1947.

RAMAN SUKUMAR

(b. April 3, 1955, Madras [now Chennai], India)

Raman Sukumar is an Indian ecologist best known for his work on the behaviour of Asian elephants and how their presence has affected both human and natural environments.

Gifford Pinchot. Harris & Ewing Collection/Library of Congress, Washington, D.C. (Digital File Number: LC-DIG-hec-21097)

As a child growing up in Madras, Sukumar was dubbed *vanavasi* (the Tamil word for "forest dweller") by his grandmother. It was during his secondary-school years that Sukumar first developed an interest in the field of conservation. He graduated from the University of Madras with bachelor's (1977) and master's (1979) degrees in botany. In 1979 he began studies for his doctoral thesis at the Indian Institute of Science, where he focused on the ecological conflict that occurs when elephants and humans use the same land. He received his doctorate in ecology in 1985 and became a professor at the Centre for Ecological Sciences, attached to the Indian Institute of Science, in 1986. Sukumar later became a Fulbright scholar and completed a postdoctoral fellowship (1991–92) at Princeton University.

In an effort to provide a safe habitat for elephants, Sukumar carried out surveys and tried to establish protected corridors so that elephant herds could move from one reserve to another. He experimented with various forms of fences around village perimeters to keep the animals away from crops and human habitation. Sukumar also helped design the Nilgiri Biosphere Reserve, the first of its kind in India, which was established in 1986. There he conducted research on climate change, tropical forests, and wildlife conservation.

In 1993 Sukumar became a member of the Project Elephant Steering Committee, which provided technical support and advice on matters of elephant conservation to the Indian government. He chaired the Asian Elephant Specialist Group of the World Conservation Union from 1997 to 2004. Sukumar also served as director of the Asian Elephant Research and Conservation Centre, a special division of the Asian Nature Conservation Foundation, an independent organization that he had helped to establish in 1997. The foundation worked closely with many governmental and nongovernmental agencies in the region to determine how to best conserve elephant habitat and manage human-elephant conflict. He published several notable texts on elephants, including *Elephant Days and Nights: Ten Years with the Indian Elephant* (1994) and *The Living Elephants: Evolutionary Ecology, Behavior and Conservation* (2003).

Sukumar's accolades included the Presidential Award of the Chicago Zoological Society in 1989 and the Order of the Golden Ark in 1997. He became a fellow of the Indian Academy of Sciences in 2000 and was inducted into the Indian National Science Academy in 2004. Sukumar was presented with the Whitley Gold Award in 2003 and with the International Cosmos Prize in 2006.

CONCLUSION

Since life began on Earth more than 3.5 billion years ago, species have adapted to changes in their environment. Most of the time, environmental change has been gradual, and species adapted to gradual fluctuations in climate, challenges from other species, or other modifications to their physical or biological environment. Extinction rates during periods of gradual change were low, as species continued to develop mutually beneficial partnerships, competitive relationships, predatory connections, or parasitic attachments with one another. Extinction rates were higher, however, during periods of accelerated and drastic change. Mass extinction events, such as those resulting from extraterrestrial impacts, continental rearrangement, or rapid climate change, eliminated whole families of organisms and a tremendous number of species.

With present extinction rates at or near 1,000 times the background rate, many scientists believe that Earth is in the throes of another mass extinction. Instead of

Moraine Lake, Banff National Park, Alberta, Can. © Digital Vision/Getty Images

natural forces, this time it is being brought on by humans and their activities. Several species have already succumbed to the effects of pollution, habitat destruction, overharvesting, and introductions of exotic species to their habitats. Many scientists contend that satisfying the needs and wants of a growing human population has placed huge numbers of plants, animals, and other organisms—as well whole ecosystems—in jeopardy, threatening critical ecological services (clean air, clean water, natural controls of agricultural pests, etc.) upon which human survival depends.

Not all species are created equal, but all play a part in maintaining healthy, functioning ecosystems. Many vulnerable species have long generation times, have inbred populations, consume a limited menu of food items, occur in small or isolated geographic areas, or are hunted or collected by humans. In contrast, low-risk species have short generation times, have multiple food options, occur over wide geographic areas, and remain relatively inconspicuous to humans. Thus, the timely creation of a management plan that protects an important, but vulnerable, keystone species may ensure the proper functioning of the ecosystem it inhabits.

The effective conservation of Earth's biological resources requires a commitment to species protection. This commitment must take the form of increased species monitoring, so species under threat can be quickly identified and effective management plans can be drawn up to minimize population declines. This commitment must also take the form of increased ecological research to better understand the roles and contributions of individual species within the ecosystems they inhabit, as well as how humans might develop better technologies to extract the resources they need from ecosystems with minimal ecological impact.

APPENDIX

Related Concepts in Conservation and Ecology

AMENSALISM

Amensalism is the association between organisms of two different species in which one is inhibited or destroyed and the other is unaffected. There are two basic modes: competition, in which a larger or stronger organism excludes a smaller or weaker one from living space or deprives it of food, and antibiosis, in which one organism is unaffected but the other is damaged or killed by a chemical secretion.

The classic demonstration of antibiosis is the destructive effect that the bread mold *Penicillium* has upon certain bacteria. The secretion, known as penicillin, has become a potent medicine in combating bacterial infections. Some higher plants secrete substances that inhibit the growth of—or kill outright—nearby competing plants. An example is the black walnut (*Juglans nigra*), which secretes juglone, a substance that destroys many herbaceous plants within its root zone.

AUTECOLOGY

Autecology, which is also called species ecology, is the study of the interactions of an individual organism or a single species with the living and nonliving factors of its environment. Autecology is primarily experimental and deals with easily measured variables such as light, humidity, and available nutrients in an effort to understand the needs, life history, and behaviour of the organism or species.

CLIMAX COMMUNITY

In ecology, the final stage of biotic succession attainable by a plant community in an area under the environmental conditions present at a particular time is called the climax community. For example, cleared forests in the eastern United States progress from fields, to old fields (with colonizing trees and shrubs), to forests of these early colonists, and finally to climax communities of longer-lived tree species. The species composition of the climax community remains the same because all the species present successfully reproduce themselves and invading species fail to gain a foothold. Because climatic changes, ecological processes, and evolutionary processes cause changes in the environment over very long periods of time, the climax stage is not completely permanent.

EARTH DAY

Earth Day is an annual celebration honouring the achievements of the environmental movement and raising awareness of the importance of long-term ecological sustainability. Earth Day is celebrated in the United States on April 22. Throughout the rest of the world it is celebrated on either April 22 or the day the vernal equinox occurs.

In the late 1960s there was an increased awareness of environmental concerns among Americans, and the prominent environmentalist and U.S. senator Gaylord Nelson sought to galvanize the conservation movement through the creation of a national celebration. Nelson—whose efforts in Congress included the passing of legislation that protected the Appalachian Trail and the banning of the use of the pesticide DDT—hired Denis Hayes, a graduate student at Harvard University, to help organize the first Earth

Day. It took place on April 22, 1970, and was designed as an "environmental teach-in" that would educate participants in the importance of environmental conservation. The two largest gatherings occurred in Washington, D.C., where 10,000 people assembled at the Washington Monument, and in New York City, where a portion of Fifth Avenue was closed to traffic in observance of the event. Across the United States, 20 million people participated, many of them at schools, colleges, and universities. The event was instrumental in gaining support for the series of environmental legislation that passed through the U.S. Congress in the 1970s, including the Clean Air Act (1970) and the Endangered Species Act (1973).

In 1990 Hayes organized a global Earth Day, which was observed by some 200 million people in more than 140 countries. Since then, Earth Day has been international in scope. By the early 21st century, Earth Day's many activities included raising awareness about a number of growing environmental concerns, especially the threat of global warming and the need for clean renewable energy sources.

FIRE STORM

A fire storm is an episode of violent convection caused by a continuous area of intense fire and characterized by destructively violent surface indrafts. Sometimes it is accompanied by tornado-like whirls that develop as hot air from the burning fuel rises. Fire storms may be generated by particularly large wildfires, conflagrations involving multiple structures, severe industrial accidents involving the release of flammable chemicals, concentrated aerial bombardment (such as over Dresden, Ger., during World War II), or nuclear detonations. Such a fire is beyond human intervention and subsides only upon the consumption of everything combustible in the locality.

FOOD CHAIN

In ecology, a food chain is the sequence of transfers of matter and energy from organism to organism in the form of food. Food chains intertwine locally into a food web because most organisms consume more than one type of animal or plant. Plants, which convert solar energy to food by photosynthesis, are the primary food source. In a predator chain, a plant-eating animal is eaten by a flesh-eating animal. In a parasite chain, a smaller organism consumes part of a larger host and may itself be parasitized by even smaller organisms. In a saprophytic chain, microorganisms live on dead organic matter.

Because energy, in the form of heat, is lost at each step, or trophic level, chains do not normally encompass more than four or five trophic levels. People can increase the total food supply by cutting out one step in the food chain: instead of consuming animals that eat cereal grains, the people themselves consume the grains. Because the food chain is made shorter, the total amount of energy available to the final consumers is increased.

FOREST FIRE

A forest fire is an uncontrolled fire occurring in vegetation more than 6 feet (1.8 m) in height. These fires often reach the proportions of a major conflagration and are sometimes begun by combustion and heat from surface and ground fires. A big forest fire may crown—that is, spread rapidly through the topmost branches of the trees before involving undergrowth or the forest floor. As a result, violent blowups are common in forest fires, and they may assume the characteristics of a firestorm.

INDICATOR SPECIES

An indicator species is an organism—often a microorganism or a plant—that serves as a measure of the environmental conditions that exist in a given locale. For example, the presence of greasewood indicates saline soil, whereas mosses often indicate acid soil. Tubifex worms indicate oxygen-poor and stagnant water unfit to drink. The presence of certain species of plants suggests how well other species might grow in the same place.

NATIONAL PARK

A national park is an area set aside by a national government for the preservation of the natural environment. A national park may be set aside for purposes of public recreation and enjoyment or because of its historical or scientific interest. Most of the landscapes and their accompanying plants and animals in a national park are kept in their natural state. The national parks in the United States and Canada tend to focus on the protection of both land and wildlife, those in Great Britain focus mainly on the land, and those in Africa primarily exist to conserve animals. Several other countries have large areas reserved in national parks, notably Brazil, Japan, India, and Australia.

The concept of a park or nature reserve under state ownership originated in the United States in 1870, and legislation creating the world's first such park, Yellowstone National Park in Wyoming, was signed by Pres. Ulysses S. Grant in 1872. Yosemite, Sequoia, and General Grant national parks were established in 1890, and during this time the idea of protecting outstanding scenic natural areas for their own sake grew into a concept of American national policy. The national park system was expanded during the following decades, and the National Park Service was created in 1916 to administer the parks. By the early 21st century the American (U.S.) national park system administered more than 350 separate areas constituting about 85 million acres (34 million hectares). In addition to national parks, the system included national preserves, national monuments, national recreation areas, national seashores, lakeshores, historic parks and sites, parkways, scenic trails, and battlefields.

Partly inspired by the American example, movements in favour of national parks sprang up in many other countries, beginning with Canada, which established its first three national parks in the mid-1880s. Nature reserves had been maintained in Europe for centuries to protect hunting grounds for use by kings and nobles, but the establishment of modern national parks and nature reserves gained momentum only after World War I or, in some cases, after World War II. Great Britain established the administrative machinery for both national parks and nature reserves in 1949. It also began establishing national parks in India and its African colonies after the war, and this practice was continued and expanded by those new nations after reaching independence. Japan and Mexico established their first national parks in the 1930s, but interest in parks generally came later in Asia and Latin America than it did in the Anglo-American countries and Europe.

The national parks of various countries differ greatly in their effectiveness in protecting their resources. Some governments provide their park systems with large enough budgets to make possible strict enforcement of regulations; others do not. Most national parks have a built-in paradox: although they often depend for their existence on tourism stimulated by public interest in nature, the preservation of their wildlife depends on its not being molested. This paradox is usually resolved by allowing visitors to travel only within limited areas in the park. This lets them see the park while it minimizes their contact with the wildlife.

NATURE RESERVE

A nature reserve is an area set aside for the purpose of preserving certain animals, plants, or both. A nature reserve differs from a national park usually in being smaller and having as its sole purpose the protection of nature.

Endangered species are often kept in reserves, away from the hunters who brought them close to extinction. In the United States, numerous wildlife refuges have served this purpose, especially with respect to birds. Nature reserves are also numerous in Europe, India, Indonesia, and some African countries.

The origin of modern nature reserves lies in medieval times, when landowners established game preserves for the protection of animals that they hunted. The idea of protecting animals simply to keep them from dying out did not arise until the 19th century.

SYSTEMS ECOLOGY

Systems ecology is a branch of ecosystem ecology (the study of energy budgets, biogeochemical cycles, and feeding and behavioral aspects of ecological communities) that attempts to clarify the structure and function of ecosystems by means of applied mathematics, mathematical models, and computer programs. It concentrates on input and output analysis and has stimulated the development of applied ecology: the application of ecological principles to the management of natural resources, agricultural production, and problems of environmental pollution.

WILDLAND FIRE

A wildland fire is an uncontrolled fire in a forest, grassland, brushland, or land sown to crops.

Fire danger in a wildland setting varies with weather conditions: drought, heat, and wind participate in drying out the timber or other fuel, making it easier to ignite. Once a fire is burning, drought, heat, and wind all increase its intensity. Topography also affects wildland fire, which spreads quickly uphill and slowly downhill. Dried grass, leaves, and light branches are considered flash fuels because they ignite readily, and fire spreads quickly in them, often generating enough heat to ignite heavier fuels such as tree stumps, heavy limbs, and the matted duff of the forest floor. Such fuels, ordinarily slow to kindle, are difficult to extinguish. Green fuels—growing vegetation—arc not considered flammable, but an intense fire can dry out leaves and needles quickly enough to allow ready ignition. Green fuels sometimes carry a special danger: evergreens, such as pine, cedar, fir, and spruce, contain flammable oils that burst into flames when heated sufficiently by the searing drafts of a forest fire.

Tools for fighting wildland fires range from the standard equipment of urban fire departments to portable pumps, tank trucks, and earth-moving equipment. Firefighting forces specially trained to deal with wildland fires are maintained by public and private owners of forestlands. Such a force may attack a fire directly by spraying water, beating out flames, and removing vegetation at the edge of the fire to contain it behind a fire line. When the very edge is too hot to approach, a fire line is built at a safe distance, sometimes using strip burning or backfire to eliminate fuel in the path of the uncontrolled fire or to change the fire's direction or slow its progress. Backfiring is used only as a last resort.

Aircraft were first used in fighting wildland fires in California in 1919. Airplanes and helicopters are primarily used for dumping water, for observation, and occasionally for assisting in communication and transporting personnel, supplies, and equipment.

GLOSSARY

autecology The study of the exchanges of a single organism or a individual species with the living and nonliving factors of its environment.

autotroph An organism that serves as a primary producer in a food chain.

biodiversity Quantity of plant and animal species found in a given environment.

carbon cycle Circulation through nature of carbon in the form of the simple element and its compounds.

circadian rhythm Inherent cycle of approximately 24 hours in length that appears to control or initiate various biological processes, including sleep, wakefulness, and digestive and hormonal activity.

decurved Curving downward.

deoxyribonucleic acid (DNA) A nucleic acid that contains the genetic instructions used in the development and functioning of all known living organisms and some viruses.

diapause Unprompted disturbance of the growth of certain animals, marked by reduced dormancy.

dormancy State of decreased metabolic activity assumed by many organisms under environmental stress or when stressful conditions are feasible.

ecology The interdisciplinary scientific study of the distribution and abundance of organisms and their interactions with their environment.

endemic Restricted or peculiar to a locality or region.

exclusive economic zone (EEZ) Area beyond a coastal country's territorial waters where it may exploit and regulate fisheries and perform various activities to its advantage.

extirpation To destroy completely; refers to local extinctions.

global warming Increase in the global average surface temperature resulting from enhancement of the greenhouse effect, primarily by air pollution.

grazer Any species that travels from prey to prey, feeding on part of each victim without completely killing it ; often used as a term to describe herbivorous species.

guild A group of species that manipulates the same kinds of resources in similar ways.

heterothermy A transitional state between cold-bloodedness and warm-bloodedness.

heterotroph An organism that consumes other organisms in a food chain.

introduced species A species either accidentally or deliberately introduced by human actions into places beyond its natural geographical reach.

keystone A species that has an excessively sizeable effect on the biological community in which it occurs.

microhabitat Refers to the conditions and organisms in the immediate vicinity of a plant or animal.

mimicry Similarity between organisms that confers a survival advantage on one.

parasitism Relationship between two species in which one obtains benefits from a host, which it usually injures.

species A group of interbreeding organisms that possess very similar characteristics.

symbiosis Any of several living arrangements between members of two different species, including commensalism, mutualism, and parasitism.

synecology Study of a population or community of organisms and their relationships to one another as well as their mutual environment.

taxonomy The classification of organisms in an ordered way that highlights natural relationships.

terrestrial Relating to the land as distinct from air or water.

torpor A state of decreased body temperature and metabolic activity assumed by many animals because of unfavourable environmental circumstances.

BIBLIOGRAPHY

CONSERVATION

Martha J. Groom, Gary K. Meffe, and C. Ronald Carroll, *Principles of Conservation Biology*, 3rd ed. (2006); and Richard B. Primack, *Essentials of Conservation Biology*, 4th ed. (2006), are introductions to the subject. Michael E. Soulé (ed.), *Conservation Biology: The Science of Scarcity and Diversity* (1986); and David Western and Mary C. Pearl (eds.), *Conservation for the Twenty-first Century* (1989), compile the ideas and experiences of several dozen leading conservation scientists.

Stuart L. Pimm, *The World According to Pimm: A Scientist Audits the Earth* (2001; also published as *A Scientist Audits the Earth*, 2004), gives an overview of the state of human activity on Earth's terrestrial, freshwater, and marine regions and discusses the effects of this activity on biodiversity. Stuart L. Pimm et al., "The Future of Biodiversity," *Science*, 269(5222):347–350 (July 21, 1995), estimates past, present, and future rates of extinction. Extensive tables and figures on the distribution and status of biodiversity are provided in Brian Groombridge and Martin D. Jenkins, *World Atlas of Biodiversity: Earth's Living Resources in the 21st Century* (2002). N.J. Collar, M.J. Crosby, and A.J. Stattersfield, *Birds to Watch 2: The World List of Threatened Birds* (1994), describes the fate of the roughly 1,100 species of birds that are globally threatened; because birds are particularly well-known, the book represents the most complete discussion of species threatened with extinction and the underlying causes of the threat. Sandra Postel, *Last Oasis: Facing Water Scarcity* (1992); and Norman Myers, *The Primary Source: Tropical Forests and Our Future*, 2nd ed. (1992), analyze the fate of freshwater ecosystems and of tropical forests, respectively. The state of fish stocks within the United States and its territories is evaluated in National

Marine Fisheries Service, *Report to Congress: Status of Fisheries of the United States* (annual). Callum M. Roberts et al., "Marine Biodiversity Hotspots and Conservation Priorities for Tropical Reefs," *Science*, 295(5558):1280–84 (Feb. 15, 2002), maps out the biodiversity of coral reefs and the factors involved in their harm.

Robert L. Peters and Thomas E. Lovejoy (eds.), *Global Warming and Biological Diversity* (1992), explores the likely consequences of global warming on biodiversity. The damage to marine ecosystems by overfishing receives major treatment in Mark Kurlansky, *Cod: A Biography of the Fish that Changed the World* (1997); and Carl Safina, *Song for the Blue Ocean: Encounters Along the World's Coasts and Beneath the Sea* (1998). Daniel Simberloff, Don C. Schmitz, and Tom C. Brown (eds.), *Strangers in Paradise: Impact and Management of Nonindigenous Species in Florida* (1997); and Stuart L. Pimm, "Species that Need No Introduction," *Yearbook of Science and the Future*, pp. 200–219 (1994), are reviews of introduced species and their role in reducing biodiversity.

Rosie Woodroffe and Joshua R. Ginsberg, "Edge Effects and the Extinction of Populations Inside Protected Areas," *Science*, 280(5372):2126–28 (June 26, 1998), shows that the foraging behaviour of some large predators, such as African wild dogs, puts them at particular risk of local extinction. Ronald L. Westemeier et al., "Tracking the Long-Term Decline and Recovery of an Isolated Population," *Science*, 282(5394):1695–98 (Nov. 27, 1998), demonstrates that the decline of genetically inbred small populations can be reversed by introducing individuals from genetically diverse larger populations.

Stuart L. Pimm and John H. Lawton, "Planning for Biodiversity," *Science*, 279(5359):2068–69 (March 27, 1998), reviews technical approaches to selecting the economically and ecologically most-effective areas to be protected for their biodiversity. William K. Stevens, *Miracle Under the Oaks: The Revival of Nature in America* (1995), describes the efforts to restore prairies in the American Midwest.

Robert Costanza et al., "The Value of the World's Ecosystem Services and Natural Capital," *Nature*, 387(6630):253–260 (May 15, 1997), calculates the annual value of the ecosystem services provided by the environment. Norman Myers and Jennifer Kent, *Perverse Subsidies: How Tax Dollars Can Undercut the Environment and the Economy* (2001); Geoffrey Heal, *Nature and the Marketplace: Capturing the Value of Ecosystem Services* (2000); and Gretchen C. Daily and Katherine Ellison, *The New Economy of Nature: The Quest to Make Conservation Profitable* (2002), discuss the economic factors that harm biodiversity and the development of economic tools to save it. National Research Council (U.S.), Committee on Scientific Issues in the Endangered Species Act, *Science and the Endangered Species Act* (1995), examines the factors involved in U.S. law in protecting endangered species.

ECOLOGY

Overviews are provided in Michael Begon, Colin R. Townsend, and John L. Harper, *Ecology: From Individuals to Ecosystems*, 4th ed. (2006); Robert E. Ricklefs and Gary L. Miller, *Ecology*, 4th ed. (1999); Robert E. Ricklefs, *The Economy of Nature*, 5th ed. (2001); Charles J. Krebs, *Ecology: The Experimental Analysis of Distribution and Abundance*, 5th ed. (2001); and David T. Krohne, *General Ecology*, 2nd ed. (2001). W.C. Allee et al., *Principles of Animal Ecology* (1949), is a dated but still important sourcebook that includes an excellent early history of ecology.

Robert Leo Smith and Thomas M. Smith, *Ecology & Field Biology*, 6th ed. (2001), offers a comprehensive survey of all aspects of ecology. Norman Myers and Jennifer

Kent (eds.), *The New Atlas of Planet Management* (2005), thoroughly surveys the state of the environment. Peter J. Bowler, *The Fontana History of the Environmental Sciences* (1992, reissued as *The Earth Encompassed*, 2000; also published as *The Norton History of the Environmental Sciences*, 1993), is a comprehensive survey from the Greeks to the present. Donald Worster, *Nature's Economy: A History of Ecological Ideas*, 2nd ed. (1994), explores changing ideas about ecology from the 18th century to the present.

Eric R. Pianka, *Evolutionary Ecology*, 5th ed. (1994), provides a clear treatment of the basics of population ecology. Peter W. Price, *Insect Ecology*, 2nd ed. (1984), is a comprehensive text on the ecology of the most diverse group of organisms on Earth. Two books that examine the dynamics of plant populations and the evolution and ecology of plant life histories are Michael J. Crawley (ed.), *Plant Ecology* (1986); and Jonathan W. Silvertown and Jonathan Lovett Doust, *Introduction to Plant Population Biology*, 3rd ed. (1993). Michael Begon and Martin Mortimer, *Population Ecology: A Unified Study of Animals and Plants* (1986), examines the range of ecological and genetic factors that influence the structure, growth, and dynamics of populations. Thorough treatments of the theory of how genetic variation arises within populations and changes over time can be found in Daniel L. Hartl and Andrew G. Clark, *Principles of Population Genetics*, 2nd ed. (1989); and John Maynard Smith, *Evolutionary Genetics* (1989). One of the most detailed studies ever made of the population ecology and genetics of a vertebrate species is reported in B. Rosemary Grant and Peter R. Grant, *Evolutionary Dynamics of a Natural Population: The Large Cactus Finch of the Galápagos* (1989). Leslie A. Real (ed.), *Ecological Genetics* (1994), discusses the ways in which ecological conditions influence the genetic and spatial structure of populations and the life histories of organisms. Ian Newton (ed.),

Lifetime Reproduction in Birds (1989), is a collection of long-term studies on the life histories and population ecology of birds. Richard B. Primack, *Essentials of Conservation Biology* (1993); and Michael E. Soulé (ed.), *Conservation Biology: The Science of Scarcity and Diversity* (1986), apply the principles of population genetics, population ecology, and evolutionary biology to the conservation of populations.

J.A. Drake *et al.* (eds.), *Biological Invasions: A Global Perspective* (1989), contains articles on the growth and dynamics of populations of species introduced into areas in which they are not native. Paul R. Ehrlich and Anne H. Ehrlich, *The Population Explosion* (1990), examines the problems of overpopulation and also offers solutions. Michael Gilpin and Ilkka Hanski (eds.), *Metapopulation Dynamics: Empirical and Theoretical Investigations* (1991), explores the importance of metapopulation structure to population dynamics and to interactions between species.

COMMUNITY ECOLOGY

Joseph Cone, *Fire Under the Sea* (1991), discusscs the discovery of deep-sea vents and the unique communities that have evolved around them. Joel E. Cohen, Frédéric Briand, and Charles M. Newman, *Community Food Webs* (1990), provides a technical and mathematical analysis of the structure of food webs and of the general patterns that have been observed in 113 previously published food webs for a wide variety of biological communities. Stuart L. Pimm, *The Balance of Nature?: Ecological Issues in the Conservation of Species and Communities* (1991), thoroughly analyzes the ways in which food webs structure biological communities. Wallace Arthur, *The Green Machine: Ecology and the Balance of Nature* (1990), gives a nontechnical account of some aspects of the structure of biological communities. E.O. Wilson and Frances M. Peter (eds.),

Biodiversity (1988), evaluates the current state and future of the topic. E.O. Wilson, *The Diversity of Life* (1992), masterfully recounts the origin of biodiversity, its maintenance within biological communities, and the threats posed by current human activities. John Terborgh, *Where Have All the Birds Gone?: Essays on the Biology and Conservation of Birds that Migrate to the American Tropics* (1989), discusses how bird population and community ecology are affected by migration over large geographic scales and by the wholesale deforestation of temperate and tropical habitats. Robert E. Ricklefs and Dolph Schluter (eds.), *Species Diversity in Ecological Communities: Historical and Geographical Perspectives* (1993), explores the ecological bases for worldwide patterns in the diversity of species.

Henry F. Howe and Lynn C. Westley, *Ecological Relationships of Plants and Animals* (1988); and Warren G. Abrahamson (ed.), *Plant-Animal Interactions* (1989), provide good introductory treatments of the variety of ways in which plants and animals interact. A more advanced treatment of the subject may be found in Peter W. Price et al. (eds.), *Plant-Animal Interactions: Evolutionary Ecology in Tropical and Temperate Regions* (1991).

Friedrich G. Barth, *Insects and Flowers: The Biology of a Partnership* (1985; originally published in German, 1982), examines the wide variety of ways in which plants have evolved to be pollinated by insects and insects have evolved to exploit the resources offered by flowers. Vernon Ahmadjian and Surindar Paracer, *Symbiosis: An Introduction to Biological Associations* (1986); and A.E. Douglas, *Symbiotic Interactions* (1994), provide nontechnical treatments of some of the many ways in which various species have evolved in association with other species and how they have developed intimate and complex relationships. John N. Thompson, *The Coevolutionary Process* (1994), comprehensively treats the specialization and coevolution of species. Peter R. Grant, *Ecology and Evolution of Darwin's Finches* (1986), summarizes the long-term research on how Galápagos finches have radiated from a single species into many species with different niches within the biological communities on these isolated islands.

DORMANCY

Dormancy and related mechanisms are treated in Charles P. Lyman and Albert R. Dawe (eds.), *Mammalian Hibernation* (1960), examining hibernation and hypothermia in birds and mammals, including articles on such thermally unstable forms as bears; Charles Kayser, *The Physiology of Natural Hibernation* (1961), an intensive look at hibernation in birds and mammals, including hypothermia and estivation, with emphasis on functional changes; John P. Hannon and Eleanor Viereck, *Comparative Physiology of Temperature Regulation* (1962), dealing with temperature regulation in both cold- and warm-blooded animals, including hypothermia and hibernation; William V. Mayer, *Hibernation* (1964), a brief, popular treatment of both birds and mammals; Kenneth C. Fisher (ed.), *Mammalian Hibernation* (1967), symposium papers dealing directly or indirectly with the topic of mammalian hibernation accompanied by over 1,300 bibliographical references relative to this topic; X.J. Musacchia and Joseph F. Saunders (eds.), *Depressed Metabolism* (1969), a variety of studies relative to hibernation, hypothermia, and thermic instability; and Harold W. Woolhouse (ed.), *Dormancy and Survival* (1969), a symposium incorporating a wide variety of review articles covering dormancy in a broad spectrum of organisms, from bacteria through mammals.

Index

A

B

D

E

F

G

H

I

N

O

P

R

S

T

U

V

W

Y